Listen to the People, Listen to the Land

Listen to the People, Listen to the Land

Jim Sinatra and Phin Murphy

MELBOURNE UNIVERSITY PRESS

MELBOURNE UNIVERSITY PRESS
PO Box 278, Carlton South, Victoria 3053, Australia
info@mup.unimelb.edu.au
www.mup.com.au

First published 1999

Designed and typeset by Phineas Murphy
Printed in Australia by Australian Print Group

National Library of Australia Cataloguing-in-Publication entry

Sinatra, Jim
    Listen to the people, listen to the land.
    ISBN 0 522 84861 3.

    1. Landscape assessment—Australia. 2. Geographical percep-
    tion—Australia. 3. Human ecology—Australia. 4. Aborigines,
    Australian—Anecdotes. 5. Australians—Anecdotes. I. Murphy,
    Phin. II. Title.
304.20994

Listen to the People
Listen to the Land

Trees are the forest
Mountains stand tall
Rivers need water
Or they're nothing at all

Oceans need fishes
Grass needs the plain
Sun must keep shining
And we all need the rain

Listen to the people
Listen to the land
What they're all saying
Can we understand

Trees are the forest
Mountains stand tall
Rivers need water
Or they're nothing at all

Oceans need fishes
Grass needs the plain
Sun must keep shining
And we all need the rain

Listen to the people
Listen to the land
What they're all saying
Can we understand

Archie Roach

Contents

Living with landscape

We turned the ignition off, and the silence of the place was eerie. The walls of the crater seemed to stand with the sole purpose of keeping the outside world from disturbing its sacred arena: a keeping place for the memories of millennia protected from the expanse of the surrounding landscape. We climbed out of our vehicle. Neither of us felt unwelcomed, but the presence of this place was intense.

We continued on the track leading to the eastern side of the crater until it abruptly stopped, then walked along a small valley—one of the many that disturb the crater's rim. We came across a waterhole a couple of hundred metres from our vehicle, but well hidden from the track. It was guarded by curved rock that had been swept smooth by wind and water since the beginning. Similar pockets of rock dot the slopes of the crater, forming countless caves and natural shelters. We had heard of rock art and hand stencils[1] in the area, but focused on being *with* this landscape rather than searching for things in it. As we climbed the loose sandstone of the inside rim, we came across several natural shelters large enough to house ten to twenty people. They had a definite sense of having been used over the years, and we then partly understood why this place was sacred to another culture and why we were not permitted to camp within its confines.

We drove out of the crater the way we had come, following the track that serpentines through the inner and outer rims shaded by desert oaks (*Allocasuarina decaisneana*) and eucalypts. Thankful for having a four-wheel-drive we traversed the plain, following a series of tracks until we found the place where we had camped on a previous trip. We dug a fire pit and lay the tarpaulin by the Mulga thickets (*Acacia aneura*) which crammed one of the numerous drainage lines that occasionally carry rain-water from the crater's slopes into the desert. After a fire-cooked meal our conversation trailed back to a sign in the crater, erected by the Northern Territory Conservation Commission and the Aboriginal Areas Protection Authority, which illustrated different world views through sharing two stories of how this landscape was formed.

Local Aboriginal belief understands that Tnorula was formed during the creation time when a large group of women danced across the sky. A mother put her baby aside, resting in its wooden baby-carrier while the women danced. The baby-carrier toppled over the edge of the dancing arena and crashed to earth, transforming into the circular rock walls of Tnorula. The mother and father searched for their child, but the infant was covered with sand and hidden from view. Today the mother, as the evening star, and the father, as the morning star, continue their search for their missing earth-bound child while the women maintain their dance across the sky as the Milky Way.

According to the Western scientific understanding of Gosses Bluff, the crater is considered one of the most significant comet impact structures by world standards. It was formed around 130 million years ago when a comet of frozen carbon dioxide, ice and dust struck the earth at extremely high velocity. The force of impact resulted in a release of energy equivalent to one million times the energy of the Hiroshima bomb, upturning sedimentary formations of 2000 metres and leaving a crater about 20 kilometres in diameter. Erosion has removed the outer rim of the crater, leaving only resistant sandstone remnants of the core near the focus of impact, which was formed when

Tnorula (Gosses Bluff)

Photograph, Ian Oswald Jacobs

deeply burned sediments were disrupted by the release of massive compressive forces (Flood 1990).

We discussed these stories in the context of trying to understand the land from different points of views, to observe the world differently from how we had been trained in our particular cultural environment. Experiencing a landscape like Tnorula and learning about distinctly different ideas of how it was created opened us to appreciate and respect landscape in a way which many people have lost through busy, complicated, modern lifestyles. Such experiences are important as they help us to become, again, more sensitive to the inherent relationship between people and land. We agreed that the value of the day's experience lay in the notion of seeing *both ways*. We didn't need the two stories of Tnorula to be printed on a bill-board to realise this. Just to walk through the landscape, aware of our own senses, helped us to respect that this landscape has always been, and no doubt will always be, a significant place for somebody else's culture.

The air chilled and the flames of the fire reduced to coals. Lying on the tarp, we gazed at a clear sky with a view of the stars that can only be experienced in the desert. The dancing women stretched from the sky's eastern horizon to the dark but unseen form of

Tnorula; at their centre was the dark shadow of the Emu, which can always be seen in the Milky Way on a dark, clear night.

The following morning began at a cool minus three degrees. After a quick meal and extinguishing the fire, we walked along the drainage line back to the crater. The dry creek led us into a valley, slowly rising from the plain as it curled between the two rims. We followed the length of the creek, which eventually climbed the inside slope. Although much of Tnorula has a strong sense of human experience this hidden valley had an untouched feeling. We continued to climb through the aroma of native mint, passing dingo tracks and patches of swept sand where wallabies had rested. We finally reached the top of the inner rim, 150 metres above the inside plain, and viewed the breath-taking panorama. We chose separate sitting rocks. The acoustics were so clear that the sound of a vehicle driving through the crater seemed too loud for the moving speck. We both watched as the car came to a halt, just long enough for its occupants to get out and photograph the waterhole before turning back the way they had come. We were silently amused at their experience, which was solely to add to the collection of places they had seen.

On this visit we spent five days getting to know the landscape, always returning to camp before Tnorula's mother began her search. The camp-fire was the focal point for discussions, as much of the day was spent in silence while we walked. We traced the valleys to the inner rim. We climbed the peaks and searched for a secure route to descend the interior slope. We walked the internal plain, seemingly sparse of vegetation under the illusion of scale. We walked around the outside perimeter of the eastern and southern slopes, searching the plains for camels and tracking their footprints in the hope of a find. We watched eagles soar in the thermals above and listened to tumbling stones dislodged by kangaroos bounding across the precarious outer slopes. We returned to our camp at the end of each day, far enough from Tnorula so as not to disturb the things we didn't understand, but close enough to enjoy its presence. We cooked and kept warm by fire and slept under the blanket of stars, the Emu always with us. We explored as much as time would allow, but we also made sure that we didn't see all.

*The importance of Tnorula as a
physical landscape lies in what
remains—its essence.*

The importance of Tnorula as a physical land-scape lies in what remains - its essence. The greatness of the former landscape lives in the timelessness of the existing one. The importance of Tnorula as a spiritual landscape lies in its power: the intense presence of place that is alive and strengthened by the unseen story of the original crater and the missing infant of searching parents. Time has eroded all but the core, and in this heart the spirit of the whole landscape lives on, enriched by the experience of new life living with the old.

The overwhelming nature of the crater is always the dominant presence, but its broader landscape is made up of smaller and more intimate 'gardens', each with a unique identity that adds to the overall character. In many ways Tnorula has been a constant source of inspiration in compiling this book. It has a clarity of definition as an Australian landscape. Its strength and power can only be attributed to time, but it also expresses fragility and a need for care to ensure that its stories will be shared in the future. In many ways this landscape captures the spirit of the Australian continent.

Our meeting with Tnorula is important to share because the raw character of place, so strong in this

The individual gardens of Tnorula contribute to the spirit of the whole landscape.

landscape, has strengthened our feeling for the connection between people and land. It has contributed to a better understanding of the stories in this book of peoples' relationships and attachment to their own personal landscapes. These stories begin in the Kimberley region in northern Western Australia, an area that for some indigenous and non-indigenous peoples holds the beginning of human interaction with the Australian landscape,[2] then transect the continent to Victoria's Western District in the southeast. The storytellers are ordinary Australians from a diversity of backgrounds, whose actions contain important messages for rural and remote areas. Although there are subtle connections between their stories, they do not hold to a consistent theme beyond the common element: landscape. The stories share activities in gardens, whether those gardens are significant cultural and historic landscapes, settlements, pastoral properties or modest home yards. They also share a combination of personal and altruistic concerns through interacting, developing or improving landscapes with limited resources and assistance, and displaying ingenuity in the face of adversity.

We feel that the stories in this book need to be shared. Like the individual gardens of Tnorula that

Much of Tnorula has a strong sense of human experience.

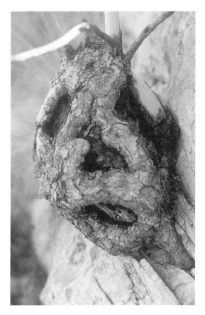

contribute to the spirit of the whole landscape, the messages of the stories form a voice about country and culture that is relevant not only to rural and remote areas but also to urban environments and their populations. The stories are personal accounts of caring for country, caring for culture, and trying to make the landscape work whilst maintaining respect. Their importance extends beyond the stories themselves to the essence of what each is saying. Combined, they hold important attitudes about cultural and environmental reconciliation, attitudes that must increasingly have an important place in our national psyche. From their humble or profound positions, the storytellers epitomise the importance of local knowledge which, through the strength of individuality, is required to forge productive yet sensitive philosophies regarding living and working with the land for mutual benefit. They each also display a spirit that reflects passion, commitment and vision, as well as a landscape ethos that is sensitive to place.

Listen to the people
Listen to the land

Image based on Gosses Bluff Impact Structure Plate 1 (Milton, Moss & Barlow 1978)

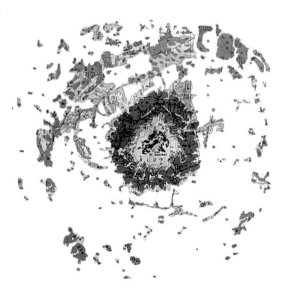

Notes

[1] Stencils are a common feature of rock art, produced by holding an object against a rock face and, with the mouth, spraying liquid pigment around it so that a negative image is created. Stencils of the hand are the most frequent, but objects such as weapons provide information about changes in material culture.

[2] Oral histories of the people of the Dampier Peninsula tell that the first people were manifested there through the Dreaming—the creative processes that gave form, life and law to the country (Goolarabooloo Association Inc. 1992). The creative ancestors then travelled to the east coast via Uluru (Ayers Rock) establishing one of the continent's major song cycles (see note 2, p. 30). According to Western understanding also, the human history of Australia is likely to have started in the northwest region more than 50 000 years ago (Flood 1995).

Black and white, a trail to understanding

Paddy Roe and Frans Hoogland

*We have to dig a bit deeper, but we settle on the surface.
We don't go to what is in our bones, that feeling.
In order to experience this, we have to walk the land. Then
we wake up to feeling, what we call le-an here, and we
become more alive, we start feeling, we become more
sensitive. And that's the time you start to experience, when
the land pulls you and takes over.
We have to learn to see again, learn to walk, to feel all these
things again. It's very hard to grasp that out of reading
books or through people talking. It's a very personal
experience.
So if there's a process where we can be guided through to
learn to get to the stage of making contact with the land
again, we get some calling of responsibility ourself.
But if you kill this country, you kill the people. We all go
down together.*

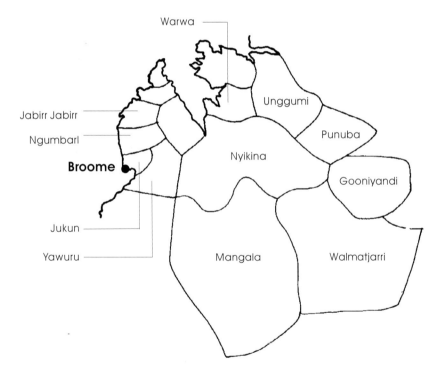

Indigenous language groups of the west Kimberley surrounding Broome

Paddy Roe was born in about 1912 on the Roebuck Plains (Pastoral) Station, approximately 25 kilometres east of Broome. Paddy is a Nyikina man, and a law-keeper of the indigenous people who live with gularabulu in what is now Western Australia, whilst maintaining the land through laws entrusted by Bugarragarra.[1] The region is part of the great song cycle of the continent, that documents the creative journeys of ancestral spirits who gave life and law to country.[2] These spirits travelled from Minyirr Jukun (Dampier Peninsula), an area of land surrounding the town of Broome (Rubibi) and the birth place of major song cycles that travel through the country to the east coast, manifesting all life and form of the living environment.[3]

Aboriginal nations in the Kimberley region maintained their culture and way of life according to Bugarragarra without interference until 1865 when the first pastoralists settled. Their coming resulted in great changes to landuse patterns, which progressively affected the Aboriginal way of life and led to suffering through introduced diseases, abduction and killing. By the time that Paddy Roe walked through Jabirr Jabirr country on the coast south of Beagle Bay, in 1931, only the elderly members of the tribe remained, as the young had been taken to missions.[4] The tribes of the Jukun people, whose country is close to Broome, and the Ngumbarl people north of Broome had already disappeared by the beginning of the twentieth century, but foresight ensured that their laws and languages were passed orally to the Jabirr Jabirr people to continue the traditional maintenance of their lands.

Paddy was accompanied on his journey by his wife, who became pregnant at Bindingankun. As they journeyed through Jabirr Jabirr country, their future family was seen by the elders as a way of continuing the law. Although Paddy was already a Nyikina lawman, he was entrusted with the custodianship of the three countries held by the Jabirr Jabirr elders. They walked him through the country teaching its names, telling him its stories, and showing him its sacred sites. Paddy extended his knowledge through learning the languages associated with each of the three countries, and the Jabirr Jabirr elders initiated him into their law. Paddy has since devoted his life to caring for the country under his custodianship.

After working for many years around the Kimberley as a station hand and windmill contractor, Paddy settled his family north of Broome in 1968 and established the Goolarabooloo community to protect the region's indigenous culture. In 1987 Paddy established the Lurujarri (coastal dunes) heritage trail to share the cultural importance of the landscape with non-Aboriginal people. He wanted to ensure the country's maintenance in the 'proper way', while the town of Broome was continuing to attract tourists and promote development. The Lurujarri trail is a parcel of land that winds 80 kilometres from Minarriny (Coulomb Point) south to Minyirr (Gantheaume Point) near Broome. It is part of a song cycle, holding the path taken by the ancestral spirits where people, animals, rocks, trees, waterholes and every component of the landscape evolved through the song of creation.

Photograph, Frans Hoogland

'I want to see country alive. This living country. Country's the big thing.'

IN the Kimberley, an Aboriginal elder sits in the shade of his tamarind tree in Broome. Paddy Roe says, 'We should all come together, European and Aboriginal people. Country man and Aboriginal man. Black and white—to look after the country'.

In his efforts to maintain his responsibility for looking after the country entrusted to his custodianship, Paddy developed a special relationship with Frans Hoogland, a Dutchman, teaching Frans tribal knowledge so that he could act as a spokesperson and mediator. This was the first step in creating cross-cultural appreciation of the land as a way of protecting it. The relationship also resulted in an invitation to other non-Aboriginal people to learn about country.

Paddy developed the Lurujarri heritage trail to teach people about the cultural importance of this coastal landscape. Walks along the trail allow people to experience the beauty of the landscape—not in the Western aesthetic sense but through an understanding of the land's power to sustain life. People from all backgrounds and cultures are learning about local Aboriginal culture and traditional perceptions of land by walking this cultural and spiritual landscape. It is Paddy's hope that the people who walk the trail learn to respect his country, and return home to become caretakers of their own country.

Paddy

Old people left the country with me. I must look after the country, that's what the old people told me. Old people didn't leave it in the writing, but everything in the proper name. Old people no read and write. I no read and write too. It's no good in writing to me. Bugarragarra never write him down anything for people. They been give me the country names: statues, everything what we got there—stones, waterholes, hills, all that country. When they hand this place over to me, they tell me, 'We not young. You got a lot of childrens coming up. So this country is yours, that's why we teach you this one, the law. You must give the name—countries, stones, hills'. I got the name for all them places, all from Bugarragarra. That's the thing [law] they been looking after too, before European

people come to their country. That's the same thing they give me. They the last old people; they hand it over to me. Must look after the country, because this has been passed over from generation to generation.

This country is right for my people. I had three young daughters, and children born, eighteen child-rens. But two daughters passed away. Young one and old one—too much drink. And I only got the last one, number two, Teresa she don't drink.

I want to look after this country, I'm thinking now. I'm sitting down under the tamarind tree over there at my place, thinking what I gonna do about the country. How we can look after this country? All right, I said, This is no good! How we gonna look after this country? I think I must do something. I must bring somebody to look after the country, I said to meself. I can't get the Aboriginal people to come out. They come out, but they like the town. All right, I said, better I get European people to help look after the country, because they're the people who got English. Anybody come, with the high English, well I can't say this language. I must get somebody with the English that can turn things around to the government. That

'That's why I teach Frans.'

was one of my ideas, one man and this tree talking. But tamarind tree never give me any answer.

What made me bring the European people more closer was because I seen what they doing in other countries. Well, I never seen it but I hear. But in this country, country's laying belly up for anybody. Not like this fighting for country, no—land's waiting for people. Country want to be safe too. Somebody gotta look after it because he not gonna be destroyed by somebody.

The first man I found to help is this man, Frans. Well that's what I was thinking. I can't look after this country with my people, so I bring this man and I been teach him to look after the country, because I want to save country. Country gotta lot of things in it. It's the law, you know, this song cycle is on the coast. That's why I want to teach my young fellas.

Frans

This is *living country*.[5] We've got to hold that one, maintain it. In order to keep country alive, you have to experience it, you have to get the feeling for it, and when you get the feeling for it and are reading the country, you can help to keep it alive. You can communicate with it. Unless you can communicate with it you won't be able to help keep it alive. See, it's like what Paddy says. You have to dig a bit deeper till you get the black soil, inside yourself. Doesn't mean you're black inside. By saying the black soil, he means the essence, like an ebony tree. On the outside it is white, but inside, the core is black. We have to dig a bit deeper, but we settle on the surface. We don't go to what is in our bones, that feeling.

When you leave camp and walk out bush the first thing to do is look for food and water. You know where to look from a feeling. You might pick up a rock and that rock has been used by people for thousands of years. Thousands of hands have rubbed that rock, and he now holds the stories. The rock, he speaks to you because there has been direct communication between that rock and people. And then you walk straight to a tree, and that tree has honey. But how did you know that tree has food? That rock, he tell you.

'See the face in that cloud, the face in that tree, the rock you can see him talking.'

In order to experience this, we have to walk the land. At a certain time for everybody, the land will take over. The land will take that person. You think you're following something, but the land is actually pulling you. When the land start pulling you, you're not even aware you're walking—you're off, you're gone. When you experience this, it's like a shift of your reality. You start seeing things you never seen before. I mean, you're trained one way or other and you actually look through that upbringing at the land. You project through your training process the reading in the land. And all of a sudden it doesn't fit anything. Then something comes out of the land, guides you. It can be a tree, a rock, a face in the sand, or a bird.

You might follow the eagle flying, and the eagle might go somewhere. Through the eagle you can see the red cliffs. Then another thing might grab your attention, and before you know it there's a path created that is connected to you. It belongs to you, and that is the way you start to communicate with the land, through your path experiences. And that path brings you right back to yourself. You become very aware about yourself. You start to tune finer and finer. Then you become aware that when you're walking the path, it's coming out of you—you are connected to it.

See, you are that land, and the land is you. There's no difference. It's hard to see the difference between nature and yourself. We have separated from it because we are told it is separate. We made a division between the garden and people. We put people on top. We have people and then everything else. So people got separated from nature and don't see themselves as a part of nature anymore. But we are part of it. Like the fish, like the birds, like the rocks, we all have our function. The land is there and is happy hearing the sounds of people. It is used to the sounds of people. It is used to the smell of people, but because we separate ourselves, it becomes lonely. The land is lonely without people, because the land, with all its forms, developed simultaneously. It's like a garden without a gardener. We are the gardeners! We are all connected. We all come from the same lifegiving force, that lifegiving essence. In the beginning, he splitting up all the time, he become rock and land simultaneously. It is not just, 'then there was this, and then people came'. No! People and everything

Photograph, Frans Hoogland

Dancing pandanus

came simultaneously. Not after, not before, but together.

We don't see the connectedness of all things. We put all the birds into a box—they are birds. We put all the trees in a box—they are trees. We put all the rocks in a box—they are rocks. But they are one and we are a part of it. We all make up the *living country*.

Country is underneath us all the time, but it's all covered up and we in our minds are all covered up. So when we walk in the land, we can't see anything for a while. We got all our possessions with us, and through these things we look at the land. Do you feel the sand you walk on? Are you aware of where your feet step? Are you aware of the trees you just passed, the birds that just landed? How much do you see? That has to shift and as soon as it does, we get a shift in mind which drops down to feeling. Then we wake up to feeling, what we call *le-an* here, and we become more alive, we start feeling, we become more sensitive. You start to read the country. Then all of a sudden there's an opening down there. Before there was only a wall, but now that tree has meaning, now that rock has meaning and all of a sudden that thing takes you. You just follow. Then you wake up, and you see a lot of things and the country starts living for you. Everything is based on that feeling *le-an*, seeing through that feeling.

I give you an example. I wake up in the morning, and as soon as I wake up I look around, I stand up and I'm gone, I'm off. I don't know why I am going. I wake up and I never do that, I usually have cup of tea first. I wake up in this place and I move straight to the reef. I'm walking on this reef and I find myself in the middle of all these little shells, these little mussels, all these little ones sticking out. And when you got cold feet, when your feet still cold, that hurts like hell; you can hardly walk on it. Before I know, I'm in the middle of it. I can't go this way, I can't go that way, I can't go back. I'm right in the guts of it. I have to walk on it. Now why the hell would I walk there. That's stupid, eh? I got rotten foot already; now I got two rotten feet.

So here I'm walking now, and I wonder how the hell did I get here? All of a sudden, right down there, there is this head sticking out of the water. Big turtle looking at me! So I go to him, and there's this hole in the rock, and the turtle is just as big as the hole and

'Are you aware of the trees you just passed?'

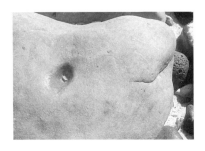

'Feeling took me to the turtle.'

he can't move. Well if he was afraid I was going to kill him, he wouldn't put his head up to let me know he was there, eh?

Well, he know I'm not going to kill him, but he needs help, he's singing out. So I go and sit next to him. He's too big for me to get him out. I sit next to him and he puts his head out again and he look at me. So I tell him, 'All right old man, old woman, old one, I try and get you and put you in the deep water, but don't bloody hit me, and be careful because I'm pretty unbalanced here!'

First time I go down, I grab him and only get him by the neck, so I have to let him go again. So he freak out a bit and he do all this thing [flapping flippers] and next minute there's all sand in the water. I can't even see him, he cover himself up. I have to wait five to ten minutes. So water clear and he come up again, and this time no problem. I hold him there and I hold him here and he moves and then I get him, and I get my balance and I tell him, 'Now one time you get up and I get up'. And we both go wop, aahh! I got him like this, and he's just hanging in there and I put him down in the water, and he go right in the water, he go down, he go down.

I'm standing in there watching him, and then he come up and he put his head up and he turn to me last time, and all right, he's off. Then from that moment I don't worry about no more shells, no more nothing. I'm just off one time. No more pain, nothing. And it seems I jump from there straight to this camp-fire and next minute I sit here and have cup of tea. Well that's *le-an*. The feeling took me to the turtle because it needed some help. And that's the time you start to experience, when the land pulls you and takes over.

Our culture, European, Anglo-Saxon culture, we not living with the land. We living from it. We taking from it all the time. We don't give back to it. But traditional people give back to it, look after it. By living there they maintain it. We dig it out. Uranium is something special inside, something special to the land. You look where uranium is. Most of the places where uranium is you find the most fabulous landscape. Extremely beautiful! That uranium is maybe like your liver or your fat or something like that. It's inside us, and it's inside the land, inside, what make our skin look

*The black sands along the trail
are rich in minerals: 'Our function
was to look after it'.*

good and our hair look good and our eyes shine. Now you start digging that away, the land gets dull, and you take our guts away we get dull too. Same thing. Nobody own the land; our function was to look after it. We're only passing through, and while you're passing through you've got a function to maintain it. That is something we have to learn: to plan for generations ahead, not just for today. We grab at it, and we leave the next generation to sort it out. What kind of people are we? Only *living country* gives you life. You destroy the country and there's nothing you can get from it.

Indigenous people are the land, the land is in their bodies. They don't see any separation from it. So the Aboriginal perspective is to maintain this country, the law, Bugarragarra—the Dreaming that gave life. These people keep life going in this country because they look after the country. We're lucky we still got this culture here because we can learn from it, we can learn how to maintain it for our future. And when we participate as non-Aboriginal people, we become more aware about ourself, how we function and how we communicate with the land. When we get to

As custodian, Paddy is trying to maintain this living country.

Photograph, Frans Hoogland

that stage where we can walk through that land, and the land has no fear of us walking through it, then we will have no fear of being in that land either. We are coming close to home then, because we should have no fear of nature.

We have a great loneliness as humans in this so-called emptiness. We are used to buildings, so we see an emptiness and it becomes a loneliness. It's like, 'I can't see no country, there are no mountains. It all looks the same to me, there's nothing there, nothing sticking out. Put a highrise building there so at least we can see something!' But when we overcome that loneliness, we realise there is nothing to be lonely about. Everything is here looking at you and communicating with you, but it takes a while to become aware of that. You can never get lost in country because it's all around you, all the time. You're with it all the time. You only get lost in your mind because you think, 'I'm a human being and that's something else'. That's how you get lost, so you look for human beings again, to get some story back. You don't listen to the land to get story back—he too alien!

Paddy clasps his hands to unite culture.

Trail walker: people are introduced to the songline through direct experience.

We have to learn to see again, learn to walk, to feel all these things again. This is why the Lurujarri Dreaming trail is so important. The Lurujarri trail will get us to listen, to start walking slowly, and to teach people. We really have to learn everything again. See, the trail is a part of a song cycle. The song cycle is been made by Bugarragarra. Bugarragarra maintains this lifeline. By cultivating that lifeline through the law, through the ceremonies of song and dance, we also maintain it. The Lurujarri trail is a part of a total song cycle and the Lurujarri custodian is Paddy Roe. Through Millinbinyarri camp, people are introduced to the song cycle through direct experience of walking and being with it, trying to understand the living quality of that country. That has to be experienced. It's very hard to grasp that out of reading books or through people talking. It's a very personal experience.

Some people might come through and have a great time and go on with their personal journey of life, not worry about this particular piece of land. But at least they have woken up to something in themselves that might be beneficial to them and to the land everywhere else they go to. If you get triggered off here to see one time, you will see everywhere. You won't lose it. When a person walks in a garden and starts loving gardening, they will see gardens everywhere. As long as he get triggered off, he gets that connection again. No matter how small it is. It's like you been in a big sleep all the time, and all of a sudden there's wind and sound. We have to wake up our senses again. If we can wake them up enough, that seeing, smelling, tasting and hearing becomes something we perceive directly. And that is the Dreaming. Dreaming is perceiving directly. When this happens, Dreaming turns into the cultivation process, the materialisation process.

The reality is that everybody is in the country, going through this country at the moment, and the danger is we are ignorant of it. We may spoil it. So if there's a process where we can be guided through to learn and get to the stage of making contact with the land again, we get some calling of responsibility ourself. It make no difference what race we are, what background we are, what kind of people we are, or what colour we are. Naturally you can't help but look after the land, because it has become some part of you too, it has given you something, woken something up. And it's not only people looking after the land; if you start looking after country, you also look after yourself. As custodian, Paddy is trying to maintain this *living country*. His idea is to have white and black walking together through this country and maintaining it and taking responsibility for it.

In this day and age, it should be very simple to say this trail is going to be here forever. We been killing so much life, we still got life here, and through some process somewhere on any kind of government level it's important that the trail be kept proper way. It's for our children. I mean in our culture we don't think further than our own stomachs. The Aboriginal culture always looks to the future. There's a duty. The people always had a duty to hand it over proper to the next generation. They learned to hand it over so life could go on and on. Why not just keep these lifelines—not only for traditional people, for everybody.

We keep on kicking ourselves out of the garden, day after day. But through the Lurujarri Dreaming trail we can all come together to feel the land, become aware of what we are feeling. You start looking after something in nature, well you're going to get life. You give water for the birds, you're going to get song, you're going to get joy around you. Simple. They will tell you story and you learn to listen. When you're here in the land, it fills you up all the time, it will always give you energy, it always make you feel all right. It gives you life. But if you kill this country, you kill the people. We all go down together. No matter what colour we are. It's a lifeline—you take your lifeline away, you take your life away.

Paddy

I teach this one, Frans. Today I shut my eye for him, and for the country. I don't have to watch him, and he's got the English. I come behind and look, sitting down, listening. That's long time, when we first started.

Now today, we climbing up. We climbing up from the ground, with the European people too,

In 1994 Paddy passed on the responsibility of custodian to his grandson Joseph Roe.

because on our own we won't be able to look after the country.

From a letter from Joseph Roe, 8 March 1999

My name is Joseph Roe. I took over from my grandfather Paddy Roe as principal custodian and law keeper for our land. The handing over was done during ceremony time according to our law and culture.

We have a Native Title Claim over our land and during the last five years we have been busy planning and negotiating with local and state governments to keep the coastal land clear of any development that would interfere with our law and culture, or damage our song cycle, places or sites.

As custodian I have to look a long way ahead. What is the future impact on this land? How many people, cars, boats and development can the country handle? And are there many social changes people bring with them?

It's important to keep the country alive like it was in the beginning so future generations can enjoy and maintain our way of life in Bugarragarra.

Besides all the politics I'm happy going bush, travelling with my mob. Hunting, fishing and teaching the young ones. That is what keeps us strong. We'd like to take all of you on the trail to see for yourselves how beautiful our country is, and to share with you our knowledge and skills, to exchange ideas about land management and our country's future.

At our community, Millinbinyarri, we have an open-house policy. People are welcome to come and spend cultural time with us. Come and walk with us.[6]

Notes

[1] Gularabulu is an area of coastal country in northern Western Australia, extending from One Arm Point in the north to La Grange in the south (Roe 1983). According to local oral history the indigenous people have been living with gularabulu since the beginning of time when the Naji spirits were embodied in the first people. Non-indigenous culture appreciates that the indigenous links to the area extend back over an estimated 25 000–30 000 years (Hoogland, F. 1999, pers. comm., 2 February). Bugarra-

garra is the Nyikina word for the Dreaming, which contains all that can be known about the process of universal creation, the origin of all things, the purpose of life, the functions of all species and their interrelationships and the journey beyond (Goolarabooloo Association Inc. 1992).

[2] The great song cycle is a group of songs that depict the life and journeys of ancestral beings or Dreamtime spirits. These travels can be recounted in long song series, in which short components relate to particular places and the entire sequence forms a map of the ancestor's journey. Knowledge of such songs shows status and power, as they are central to insight about the Dreaming and country. The song cycle from One Arm Point to La Grange has been in existence since law and culture were given to the first people, remembered today in human imprints throughout the landscape including those of Marala, a law-giver who left his markings during the time of the dinosaurs (Hoogland, F. 1999, pers. comm., 2 February). 'Country', as opposed to the non-indigenous interpretation of 'the country', refers to the landscape, the sea, and even the sky (Rose 1996). It includes the animate and the inanimate, and the interdependence of all life. Speaking of 'country' implies a spiritual attachment to traditional land, which is considered home (Nathan & Japanangka 1983). More importantly, 'country' is a part of identity as indigenous people do not belief that they came from somewhere else, but from their traditional land itself.

[3] Rubibi has immense spiritual and historical importance as a 'life-giving place' to local Aboriginal people, just as Jerusalem, Mecca and Benares have to other cultures.

[4] The practice of missionaries included collecting part-Aboriginal children for education, which often resulted in lost connection to cultural heritage and family. Young Paddy's skin colour revealed white parentage, which placed him at risk of being taken to Beagle Bay Mission. However, his tribespeople took care of him, and he was later fully initiated under Nyikina law (Roe 1983).

[5] Living country is 'where the land is whole and complete; where the interaction between people and land is alive through law and culture; where the spirit of the land is 'standing up', and vibrant.' Living country differs to cultivated country, which is 'where there is cultural confusion; the land has been replanned one layer over another; the spirit is hiding, withdrawn, waiting.' Dormant country is where there is no interaction of people with the land through their traditional law and culture; the spirit of the land is awake but resting, waiting' (pers. comm. Frans Hoogland 26 February 1999).

[6] Information about the Lurujarri trail is available through the authors, or at Millinbinyarri on telephone 08 91922959.

Moving back to country

Laurie and Penny Cox

This is where my grandmother Lena came from. The Nimanburru tribe.

We gotta find a little bit more of our backgrounds. We are starting a community of Cox and Manado families. The main reason for the movement out here is to be independent, you know, get away from any government funding.

It would be good if the kids learned something about the land and the language. We also need business projects so that when the kids grow up, at least they can come back and work out here for themselves.

To move out is to be part of the Homeland movement. That's one of the main reasons, and to make future for the kids.

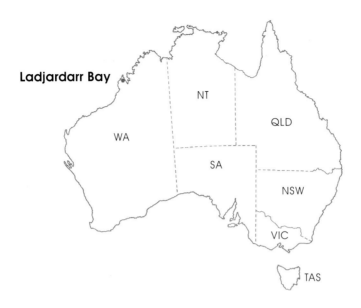

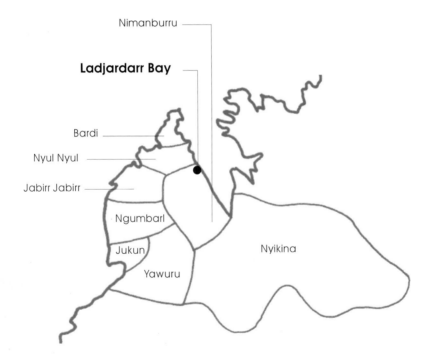

Indigenous language groups of the west Kimberley
surrounding Ladjardarr Bay and Nimanburru country

Christian missions have played a significant role in indigenous affairs throughout Australia from the early 1800s. Aboriginal people came into contact with Anglican, Methodist, Catholic and Lutheran missionaries between 1820 and 1840 in southeast and southwest Australia. During the second half of the century, they became resident at more permanent mission communities such as New Norcia in Western Australia, Poonindie and Point Pierce in South Australia, and Lake Tyers in Victoria. With islanders, they came into contact with missionaries in the Torres Strait from 1871, followed by establishments at Yarrabah, Hopevale and Mapoon in Queensland, and the Daly River in the Northern Territory.

The early missionaries failed to evangelise and 'civilise' Aboriginal people beyond teaching basic literacy to children and basic knowledge of the Bible. As the role of the missions became more established, their success in converting people was limited by adoption of the Christian faith by Aboriginal people as an extension, rather than replacement, of their own beliefs. European culture continued to take control over larger areas of the continent, and the missions became active in both the preservation and destruction of traditional indigenous culture. Some missions supported cultural maintenance, while

St Mary's church in Beagle Bay

teaching skills valued by non-Aboriginal society and providing a sanctuary from both white and traditional Aboriginal enemies. Others strove to suppress many aspects of culture by imposing Western values at the expense of traditional customs. They banned language, song, dance and traditional ceremonies such as initiation, which are essential means of passing on cultural heritage; they contributed to the erosion of customary authority by separating family members and appropriating authority from elders; they exercised a wide control over Aboriginal people's lives; and over time they fostered dependence and became paternalistic institutions of welfare.

The first mission in northern Western Australia was established by the French Trappist monks. It was developed on Nimanburru land at Disaster Bay in 1885, and later shifted to Nyul Nyul land at Beagle Bay on the western side of the Dampier Peninsula. The Trappist monks maintained the mission at Beagle Bay from 1890 until 1898, when they were withdrawn. They were replaced by the German Pallottine Order in 1901, who remained in control of the community until 1978, when management was transferred to the Beagle Bay Aboriginal Council.

In the spirit of the outstation, or homeland, movement[1] the Cox and Manado families decided to move from Beagle Bay Community to re-establish their cultural heritage and links with their country. The decision to leave the community was based on the wish of Laurie Cox's grandfather, Jerome Manado, to see his family move to the country of their ancestors, the Nimanburru people. Discussions with Beagle Bay Council for a sublease of land 60 kilometres east of the Beagle Bay Community took place in 1989, and the Ladjardarr Bay Corporation was incorporated in 1990. In 1991 the Cox and Manado families began their move.

The shift to Ladjardarr Bay (Disaster Bay) not only enables the families to be reunited with their country but also allows them to create a future for their children through establishing economic enterprises. The Ladjardarr Bay Aboriginal Corporation has created a five-year development plan, which includes landscaping, horticulture, a building training program, and aquaculture and tourism ventures.

Laurie and Penny Cox and family have moved from Beagle Bay Community back to their ancestral lands. This move allows a new future for themselves and their children. Both Laurie and Penny grew up in the Beagle Bay Mission; they married and began to raise a family. They remained in Beagle Bay long after management was transferred from the mission's Pallottine Order to the Beagle Bay Aboriginal Council. Their decision to help establish a new community at Ladjardarr Bay, the country of Laurie's ancestors, in 1991 was a response not only to his grandfather's wish but also to their questioning of whether Beagle Bay was the best environment to raise a family. The Coxes are committed to family and culture, and being reunited with their country has enabled them to work against the odds that recent Australian history has dictated towards securing the family's cultural and economic future.

Laurie

Yeah, this family country. This area was where the first mission was set up in the Kimberley. The Trappist priest came and gathered the old people from around here and they started up a community. I don't know what happened, but they moved out from here and they took everyone along with them. They went around the coast way, picked up everyone and moved down Lombadina way. This is where my grandmother, Lena, came from. The Nimanburru tribe. She'd been brought away from here with her mother and grandmother to Lombadina and then to Beagle Bay Mission when she was a little girl. Those were the only family from here, and we told her we would come back.

Me other old *lulu* [grandfather], Jerome Manado, he was born further up the track there. He was born up at Gurramulla, Goodenough Bay side of Cornambie Point. When they went to Beagle Bay, they send him back to Lombadina and he grew up there. He is still alive. That's why he wanted us to come back here—start something while he was still alive. Ah yeah, it's important.

The only one who knows about the history is an old bloke up at Lombadina and he should be out here soon. Gotta pick him up. Old Sandy Paddy, a Bardi tribe elder and law-keeper of the area. He was going to come out here and do a little bit more things, law things, because he's a part of the family too. I don't think any of our family knows much about Aboriginal laws and culture. We weren't brought up to know anything like that because of the mission. We gotta find a little bit more of our backgrounds.

We are starting a community of Cox and Manado families, plus a few others who helped us from the start. This place had been on the drawing-board before we moved and we spent one year getting everything organised to come out here from Beagle Bay Community. We have been living here since May 1991. For the past year we been living in caravans. See, we came out with nothing at all.

I think funding is one of the main things. You can do all the groundwork on your own, but you gotta have funding to work along in the beginning. We can use government money at the moment; we are entitled to it and we might as well get funding while it's still available. But I think ATSIC [Aboriginal and Torres Strait Islander Commission] lacks people who want to go ahead and do something for themselves, you know? See, the main reason for the movement out here is to be independent, you know, get away from any government funding or anything like that. The moment we can stand on our own two feet, we will.

The hardest thing about moving is beginning all over again, getting organised. Uncle Teddo was out here for about three weeks before I came out. He just

started up a little shed—that's all. When I came over we cleared the whole area. That went on for about three months, three to four months. We had no power for the first five months. We had been carting water for four months from 21 kilometres away with a little pump, every afternoon, just with water cans, five or six water cans. Four to five months after being here we got the bore and then we started to grow plants in the garden. It was sort of like coming into another world when we left Beagle Bay. You'd sort of relax— you can sit outside without somebody coming up arguing, you can have your full-time sleep at night. Well, in Beagle Bay you couldn't do that—I couldn't do that.

Penny

Me and Laurie used to sleep in the four-man tent and the kids used to sleep in the Toyota. We had our own Toyota. Or sometimes we used to sleep outside in the open, just on the ground. We had just a fire, a little camp-fire going. I find it really good, being out here when there's no light. Sit down in the quiet, you know, listen to the sound of the bush and the sea and the birds. No shouting or screaming or no cars running around all over the place. At first I used to come up and down on the weekends and bring the kids. We finally gave up the house in Beagle Bay and came out here to live. It took about eighteen months for the whole family to move.

We got a reliable community vehicle to take the kids in and out to school, a one-hour trip each way to Beagle Bay. There is nothing in Beagle Bay for them any more besides school. With so many kids, it is harder living here than in town. The kids was our main problem. They didn't like moving! Now they know it's their home.

Laurie

We had to drum it into them for a while, to try and get them to learn the reason why we left Beagle Bay, and now, when Denise, our oldest daughter, would come back from school in Perth, she would sort of drum it into the kids' heads as well. She didn't want to go back to Beagle Bay; she just wanted to go straight out here again.

Penny

At the most there are seventeen children out here [in November 1993]. From seventeen right down to one 1-year-old. There is about twelve workers and just us four womens.

Mainly all the family will be out here. When you have the families—uncles, cousins, aunties, nephews, nieces—all together, they sort of mix together well.

When we get this pre-primary school going all the little ones will stay here and they [the nuns from St Mary's in Beagle Bay] will teach them right here. But that won't be until 1994 or 1995. To start this pre-school up, maybe one of the teachers from Beagle Bay will come out here during the week and teach them till Friday, then go back. She could come here and teach the little ones or give them some work to

'I am used to the bush, I suppose —never really liked being in town. We are used to the bush, all of the time since we grew up.'

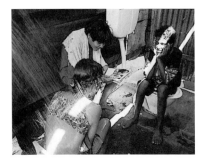

When you have the families all together, they sort of mix together well.

'We wanted Sandy to come out here to smoke the mans—smoke from koongkurra *tree'.*

do and one of us will work with her. If she is sick or something and can't come up, one of us will give the little ones work to do for that day.

Friends of our kids come out here every weekend. They rather come out here than stay in Beagle Bay. That's why Laurie wants to get more of . . . what you call them? . . . activities, like more horses, so kids can go out riding. They can go fishing, they follow kangaroo tracks, they can go out for long walks along the beach or travel on the truck, and they can spotlight crocodiles at night. If they go swimming we go out with them, see. We get in a bit—knee deep—and always looking around. You can see them crocs when you look out to the sea, but sometimes when they are along the shore they are very sly. I've seen a couple like that. You are expecting them to come from the deep, you know, and they come along from the shore.

The dogs are always with the kids when they go out. We think Chippo, the family dog, was taken by a croc. The crocs and centipedes and what else, all the other critters out here, they are just a part of it. Yeah, a part of it, not something to worry about.

It would be good if the kids learned something about the land and the language. I don't know any languages—Uncle Teddo over here, he could teach the kids something because he knows some of the language. Uncle Teddo would be the right man to do it in his spare time. Teddo was supposed to pick up Sandy Paddy from Lombadina a long time ago. Sandy was supposed to come up here and smoke the boys, but he didn't get around to doing it.

See, Laurie is not sure where the sacred grounds are. They don't know what they saw when they go over the rise. They sort of waked them spirits up. That is the main reason why we wanted Sandy to come out here to smoke them, especially the mans—smoke from *koongkurra* tree.[2] I don't know its proper name but it will keep spirits away. It won't be Laurie that will get sick or anything. The spirits will turn back to the wife or one of his kids. Laurie wants to protect his family.

You see, on my side of the family, my mother's grandmothers and ancestors, I wasn't told about laws and the like. My mother and father never told me anything. I don't know, maybe I never asked them.

Some womens today can't handle their kids, like how my sister couldn't handle Bodine; they sent her to us to look after. Well, that's what happened to us. I was born in Broome, but when I was about six I was sent to Beagle Bay with my eldest brother by my parents. Mum sent us out to Beagle Bay. I do remember when I went there, when I was small. It was run by Sisters and Brothers and that place work really well. There was no problems then. I feel bitsey [not good] about past because I don't know any culture really. I grew up with half-half, the same as Laurie. He grew up half-half: half religion and half old people belief. Laurie wants us to keep going. He wants to build a little church here later on, for Sunday. He will pick a good spot. We can keep both cultures at the same time.

We also need business projects so that when the kids grow up, at least they can come back and work out here for themselves. When they get older and come back here they can keep the garden going, and these things they are going to put up here—yabbies. For tourists, we will show them how we live and show them the bush and Goodenough Bay, the caves and the coloured rocks.

A Ghost Gum on the coast indicates an area that includes sacred grounds.

Fencing the aquaculture dam

Photograph courtesy Ladjardarr Bay Aboriginal Corporation

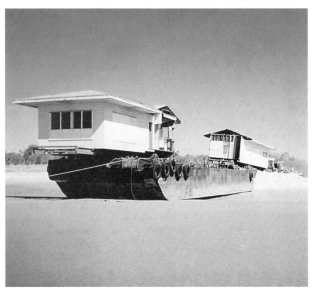

*New house transported from
Derby across King Sound to
Ladjardarr Bay, 1994.*

Laurie

We got 514 acres. We'll probably get some more land later on down the track. I think we'll probably be looking at cattle later on. Might be next year or year after. This can be another money-making [project] for our community of six or seven families. But you need a good breeding programme. You start running cattle, you gotta start fresh and new and everything. You got to start breeding your own stock.

Penny

Lawrence, our oldest son, is a home man. We will be sending Lawrence to Perth to do a course. There is going to be a man out here for twelve months doing work on that yabby thing, from the same yabby company where Lawrence is going to work down in Perth, so when the man leaves, Lawrence will take over. The young ones that come after Lawrence will be able to learn from him. He'll be the teacher.

We sent Denise to Perth to get a better education. At the time they didn't have secondary school in Beagle Bay.

Laurie

I'm glad she's going right through all her education, schooling. After going to the US for an additional year of schooling,[3] she probably will be back here and work for a while, maybe three or four years, and move out again. I think she feels she's gotta give the family back something.

To move out is to be part of the homeland movement. That's one of the main reasons, and to make future for kids. So we are out here to set up small business for the kids, for their future. I don't think unemployment will be around too long, so we are setting little businesses for them so they can carry on. We got about four projects for us in the future—aquaculture, farming, vegetable garden, clay and tourism. ATSIC's helping with a yabby project at the moment.

Photograph courtesy Jo Henry

New developments at Ladjardarr Bay since the story was written include a store, school, community office and additional housing. The Ladjardarr Bay Corporation is in the process of establishing tourist ventures, which include horse riding, crocodile spotting, fishing and camping.

Notes

[1] Outstations are small Aboriginal settlements or homeland centres, usually of people with close kinship ties, in lands significant to them for cultural and/or economic reasons. The movement of people from communities to establish outstations on their homelands began in the 1960s. The parcel of land that has been named Ladjardarr Bay is one of a number of outstations within the Beagle Bay Reserve.

[2] The Nyul Nyul name for the Conkerberry (*Carissa lanceolata*), a small spreading shrub scattered across the Dampier Peninsula and also found in the Northern Territory and Queensland. Apart from the use explained by Penny Cox, the smoke from its burning wood repels mosquitoes, and green leaves are used as a smoke medicine to cure diarrhoea in infants; the sweet fruit is used medicinally for sores and the roots for various medicinal purposes (Isaacs 1987).

[3] Denise won a Rotary International Exchange scholarship to the United States in 1993.

Yukirrinya

Nigel Carrick

What struck me in Kintore was that the build-up of rubbish benefited the vegetation, so I started to think about ways to use rubbish for protecting and mulching trees.
The whole idea is to slow water runoff, catch organic matter and catch silt. I had started to think in terms of a waste management system that was synonymous with a landcare system by using discarded material to complement tree planting.
The trees help to reduce dust-borne diseases by providing shelter, and plant uses such as firewood and carving wood for artefacts are important for culture. The increased vegetation also helps produce the biomass that can be used to attain a system of food.
Money is continually spent on bricks and mortar while people go on living outside. So I have been trying to create areas separate from houses that you can be comfortable in.
I respect people who have a long-held attachment to place as an identity. I have a sense of belonging and a respect for country.

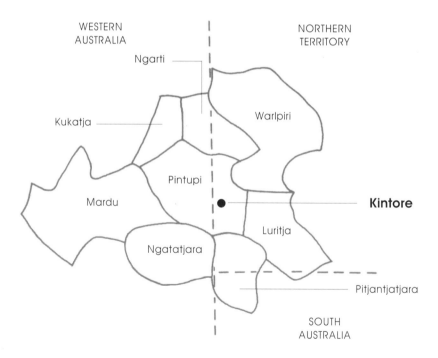

*Indigenous language groups of the central/western
desert region surrounding Kintore and Pintupi country*

Kintore is an Aboriginal community 520 kilometres west-northwest of Alice Springs in the Northern Territory. It is on the Pintupi homelands in the Gibson Desert, which extends across central and Western Australia. The region has an unpredictable rainfall with an annual average of less than 250 millimetres. The broader landscape consists of mountain outcrops, sandhills and spinifex plains with areas of Mulga (*Acacia aneura*), Desert Oak (*Allocasuarina decaisneana*) and grass country.

Before contact, Aboriginal groups living throughout arid Australia required a large area to satisfy their food and water requirements, and their territory boundaries were therefore larger than in coastal and temperate zones of the continent. It is estimated that 200 square kilometres of country was required to support one Pintupi person in essential resources for subsistence (Long 1971). In conjunction with seasonal availability of food and water, this determined a nomadic lifestyle. People relied on rockholes, claypans, soakages and wells for a constant water supply, moving to the remote areas after rain to use seasonal supplies and returning to more permanent sources as the country dried. Variations in food supply also determined people's movement. During good seasons large groups came together for organised hunts, which also allowed ceremonial gatherings to renew social and cultural ties. During the poorer seasons, game was hunted in smaller and more widely dispersed parties. The staple diet of the Pintupi people included lizard, kangaroo, emu, turkey and a variety of seeds, seasonal fruits and vegetables.

After brief contact with the early European explorers, their harsh environment protected the Pintupi people from further contact and protected their traditional lifestyle until the late 1920s and 1930s. By the late 1950s and early 1960s many had moved to settlements, missions and pastoral properties attracted by food availability during periods of severe drought. The government's policy of assimilation also facilitated the move of *yanangu[1]* into settlements. Assimilation was designed to include indigenous people as part of the broader Australian community, through them having to accept the same responsibilities, customs and beliefs as those in mainstream society. In reality, assimilation had devastating effects on Aboriginal societies through the dispersal of communities, intentional break-up of family groups, and displacement of people from land, culture and language, which resulted in lost identity. Indigenous people were to be 'bred out' through absorbing 'mixed bloods' into mainstream society and segregating the 'full bloods'— those who were not considered capable of being assimilated and would eventually die out on reserves. People were centralised in settlements around the continent, and children who were forcibly removed from their families for schooling often never re-established family contact. The missions played an important role. Although they often provided protection from settlers, pastoralists and pearlers, they also facilitated a loss of cultural heritage through firmly establishing English and Christianity at the expense of language and traditional ceremony.

Assimilation became policy in the Northern Territory in 1953. This led to many of the desert people moving to settlements such as Papunya, which was established by the government in 1959 to facilitate assimilation. The establishment of the Woomera Rocket Range in 1946, and later the construction of the Gunbarrel Highway in 1958, opened up the central desert and provided access to move people from their homelands during the 1960s. During this period the Pintupi people moved into Papunya, 260 kilometres west of Alice Springs, along with Luritja, Anmatyerre, Warlpiri and Arrernte people. Problems of internal conflict, alcoholism and riots resulted from dislocation from traditional country and mixing with other language groups. Papunya was considered a 'sad place'; a place where people were depressed through 'worrying for their country'; an overcrowded place where many died from malnutrition, disease and injury.[2]

An unofficial change in government policy from 'centralisation' to 'dispersal' created an opportunity for the Pintupi people to settle on Luritja country 80 kilometres west of Papunya at Mt Liebig in 1978. In 1980 they moved further west to Kintore on their own country, 275 kilometres west of Papunya and about 50 kilometres east of the Northern Territory–Western Australian boarder. A hand pump was provided by the Outstation Resource Centre, and in 1981 Kintore (Walungurru) was incorporated, officially marking the

Aerial views of Kintore Community and surrounding landscape, 1996

Pintupi return home. Today Kintore has over fifty houses, a fluctuating population of up to 500 people, and non-Aboriginal service providers in health, education and administration. The community also operates as a resource centre for twelve outstations throughout the Pintupi homelands.[3]

Although the Pintupi people have returned home, their now predominantly sedentary lifestyle is a legacy of government policy and recent history. Before contact, their nomadic lifestyle not only limited stress on the environment but also sustained people's physical and social well-being. Permanent settlement has resulted in environmental degradation, one of the effects of a mobile group of people 'sitting down' in one place for a long time. This in turn has led to environmental and health problems caused by the depletion of vegetative resources. The impact of vehicles, which now play a vital role in the mobility of Aboriginal society, and the loss of vegetation used for traditional purposes such as firewood, have contributed to loss of top soil, increased dust movement, and reduced shade and shelter. The predominantly outdoor lifestyle of the Pintupi, and the mismatch between the socio-cultural environment of the community and the physical environment of the settlement, have contributed to poor community health.

'The task is not so much to see what no one yet has seen, but to think what nobody yet has thought about which everybody sees.'[4]

NIGEL Carrick is Kintore's Parks and Gardens Supervisor, and is working with community members in planting trees, shrubs and vines for windbreaks and shade. Nigel's planting work is a landcare strategy that responds to the need for re-establishing vegetation lost as a result of the community's predominantly sedentary lifestyle enforced by the permanent settlement. It also aims to improve community health by increasing shelter throughout the community, as people retain their outdoor lifestyle. Establishing better health is a vital part of ensuring that Pintupi people hold on to their country and law through the maintenance of culture, language, tradition and sacred sites.

Nigel grew up in Melbourne. His early avid interest in bird watching led on to botany, plant communities, bird-nesting sites and food sources. Concerns about consumerism and a need for habitat preservation directed Nigel to work with conservation groups and recycling organisations. After travelling to Alice Springs and working as a landscape gardener, Nigel was invited to Kintore in 1987 to teach a land management course. He was introduced as Yukirrinya (green one) and worked in the community until 1995. The plantings now established throughout the settlement aim to provide sheltered communal areas for meetings, reduce soil and water erosion, establish shelter belts to reduce wind and dust, develop systems of food production, and aid septic-tank waste removal. Nigel's resourceful use of community litter as mulch not only provides rubbish with a function but has also increased the survival rate of trees planted in schemes for low maintenance.

Nigel

I don't know why people get so offended, I mean it's a prejudice. I've always gone through bins to sort rubbish for recycling. In Melbourne I would go through rubbish to find what people hadn't managed to use somewhere else. I remember doing it to my mother and she couldn't stand it. I went through her bin and I said, 'Look, there's no rubbish in there'. I have childhood memories of collecting milk-bottle tops, plastic containers and cardboard for preschool art and crafts. I have always sorted rubbish into recycling because of our constant taxing of world resources. Recycling in Kintore has its limitations as it is 550 kilometres from the nearest centre, but going to Kintore gave me the opportunity to sort and redirect waste and thereby to stop land filling at one place.

Before contact with white Australia, it was appropriate for Aboriginal people to disregard waste because everything used was bio-degradable and the people were nomadic. It is important to remember this when trying to understand why there is so much rubbish in a community such as Kintore. Most visiting white officials can not think of anything better to say in their first ten minutes other than, 'What a lot of rubbish. Why can't this be cleaned up?' DAA [Department of Aboriginal Affairs] officials even proposed D9 bulldozers to clean it all away, which would have been disastrous for the topsoil. It would make chances of revegetation, which is dependent on infrequent rains, slim, compounding the already bad dust problems within the community.

What struck me in Kintore was that the build-up of rubbish in the community benefited the vegetation. I noticed that discarded car parts protected a ring of vegetation from being driven over, and I slowly began to view rubbish in a positive light as far as topsoil was concerned. In some cases, mounds of rubbish were built up on fence lines and subsequently covered by wind-blown sand. These areas collect seeds, initially forming a green belt of grasses, ground covers, herbs and small shrubs. As more rubbish collects the dune slowly increases in width and height and gives rise to a vegetation belt. All fences in Kintore show this build-up of soil, rubbish and organic matter. The rubbish is an organically inert substrate aiding water retention and germination of seedlings after rain, whereas organically mulched beds will be termite food upon any drying out.

The build-up of discarded items is a major resource. Aboriginal communities tend to be really scarce in new resources as we get a minimal budget

These areas collect seeds, initially forming a green belt of grasses, ground covers, herbs and small shrubs.

through council from local government funding. Organic mulch has always been scarce and it's questionable to spend $120 for a bale of Redgum chips from Alice Springs, so I started to think about ways to use rubbish for protecting and mulching trees. I have also introduced species for green manure,[5] as there are no animal manures available and human excretion is not considered acceptable.

I started our Parks and Gardens team from the sanitation and garbage funding that the council gets. In the high-use areas of the community such as around the council office, store, clinic and women's centre, we ran a constant rake and weed team, collecting bales of organic litter or weedings and bales of mixed rubbish including plastic, paper, polystyrene and tin. These bales were used to mulch shelter belts and isolated trees. The Parks and Gardens team was made responsible for this as it was perceived that the trees we planted and protected with wire and fences were responsible for the wind-blown rubbish that collected in and around them.

Picking up rubbish was certainly not a priority of mine but I strove to make sure that our time supported our planting program. In an eighteen-month period we collected 120 bales of rubbish from two hectares in the high-use area of the community. However, doing this more than once a month was too time-consuming and, although we did clean-ups for visiting officials, I preferred to let it build up on the fences so that it was well worth the effort to collect. It

was far more beneficial to the community to plant trees and leave wind-blown rubbish piles for a three month clean-up. Store rubbish was also a resource of large quantities of cardboard and bales of plastic and polystyrene, and I collected fruit and vegetable organic waste for the compost heap.

The entire year of 1992 was spent mulching shelter belts with raked rubbish. After the first three months I didn't have to worry about drought stress or salination by evaporation, there were no weeds in competition and the top soil around the trees was protected. Some plots were built up with a 30 centimetre layer of rubbish and a 30 centimetre layer of weeding over the top. This was held in place by laying star pickets or bike frames over the top as winds can be 70 kilometres per hour with willy-willies that will suck up and spread any ground litter. These mulched plantings have been dug out to form a bowl so they can be flooded, and are bordered by an arrangement of scrap metal including engine blocks, axles and wheel rims to form a kerb and a bank for the planting to pond. The beds can be flooded without flowing out across roads or bare dirt where an evaporation rate of nearly 8 metres per year would waste it. The result is a maintenance-free planting that needs water once a month for the first four months and then every three months after that. After eighteen months it can be watered twice a year for optimum growth.

I started to place rocks with the fence-line trees near the council area because rocks protect trapped

The build-up of rubbish benefits the vegetation.

Photograph, Nigel Carrick

moisture in the ground. They also become a silt and organic matter trap. You can lay rock traps right across the landscape to slow water runoff, and native plants will germinate in the accumulated silt. You can then go back and put a ring of rocks around each seedling that germinated; no watering, no maintenance, just a hell of a lot of time placing rocks. That's what we did at the back of the church, and staggered them across the landscape. I then got the idea of using our resources; doing what I was doing using car rims and car tyres. The whole idea is to slow water runoff, catch organic matter and catch silt—and you can't wash away a car tyre when it is dug into the ground.

All of Kintore's land surfaces could regenerate if it wasn't for the land degradation caused by unrestricted vehicles. The vehicles compact the top soil forming ruts, and the increased water runoff contributes to a loss of its capacity to regenerate. There is nothing more devastating than to see a planting of 120 trees being continually driven over. When I observed people restricting access to roads for drunk drivers using engine blocks, I started using engine blocks along with any other heavy metal to protect the trees. In the low areas of the community there were a lot of car parts and old camp structures that could be used to build vehicular barriers. On one old road I placed semi-circles of car tyres and rims and

'Heavy-metal mulching.'

watched the grasses emerge while absolutely no plant regeneration occurred elsewhere on the hard surfaces. We also protected a planting of 120 eucalypts with car bodies. Car batteries, radiators and aluminium cans can cause some poisoning of the ground and water table over a long term and all three items have been sent to Alice Springs on a regular basis, forming part of my Parks and Gardens budget.

I had started to think in terms of a waste management system that was synonymous with a landcare system by using discarded material to complement tree planting through water harvesting, reducing evaporation and creating barriers that make people think twice about driving over trees. Many people say you can't plant trees in the desert unless you have a large amount of water, but our planting demonstrations in Kintore challenge this. The most depressing aspect of working in Kintore is trying to convince the Office of Local Government that rubbish or litter aids tree survival and that using it is a positive thing to be doing.

The loss of vegetation compounds dust and wind erosion, which has a cumulative effect on the health of community members. Tree plantings help to reduce dust-borne diseases by providing shelter, and other uses such as firewood and carving wood for artefacts [6] are important for the culture. I received a letter from the community's past doctor which stated:

> *Repairing the devastation of loss of ground
> cover plants and trees in the main living
> area of Walungurru [Kintore] will have*

*long term implications on the health of the
community. The simple presence or
absence of a shrub or tree will not in itself
be the answer, but be part of the process of
appreciating the present problems . . . Diseases such as Trachoma, chronic secre-
tory otitis media and respiratory diseases
(childhood pneumonias) are either endem-
ic or present at exceptionally high rates.
Reducing dust and dirt, thereby diminish-
ing the need for refuge and subsequent
overcrowding in what houses already exist,
would help begin to diminish their presence . . . Brushwood for the elderly needing
to live and maintain their lifestyles in traditional wilytjas must be sought further and
further away . . . A source of firewood for
cooking bush-tucker and keeping warm is
required. These people are responsible for
the maintenance of cultural integrity . . .
Returning that place to its previous state
would display a strength of culture and an
appreciation of the need and methods to
grapple with a problem at present making
its own people sick.[7]*

Trees improve environmental health by creating shaded living environments protected from wind
and dust. Aboriginal people in Kintore want to be out
of the wind while sitting in the sun during winter. In
summer they want the breeze, and they want shade
provided by a wilytja or shade tree.[8] Money is continu-

Contemporary *wilytja*

ally spent on bricks and mortar while people go on living outside. People have been observed to be sicker in and around houses than under the tin humpies they use, day and night, to get out of the wind and dust, and when the technologies in the houses break down. But no one wants to risk any money from the Aboriginal health strategy to tackle the big question of their actual living environment. These people have a permanent culture but they are quickly being robbed of their future.

There is a funding need for a nursery, and a big works team with tractors and shredders, but the council is left with counting meagre pennies that have to encompass all aspects of Pintupi life, and funding needs for trees are often overlooked.

Increased bio-mass is the only way ahead. The community set the priority of planting around the main meeting areas and places where people congregate such as the clinic, church and women's centre and we planted again and again to establish trees. One of our concerns has been to look at the areas around the council office that are used for large assemblies of people. The southern and eastern perimeters of a community meeting area were linked for a large continuous windbreak to the yard of the Papunya Tula artists' shed. Indigenous trees were planted for shade and other trees planted for windbreaks. A council meeting area was then defined by a wilytja and a line of seating rocks placed along the western edge. This area was planted to block the potentially nasty northwesterly winds.

I have been trying to create areas separate from houses that you can be comfortable in by planting trees. After planting twenty or thirty trees at every house one could see who had cared for them and who hadn't. Those people who loved their gardens got lots of special help. Mrs Bennett is one person who loves her trees as she loves her dogs. Now she has a solid windbreak on all southern and eastern fronts and she has a corrugated iron humpy and shade shelter adjacent to her house. If you looked at the amount of time she spends in her house you would realise that the house is just a set of power points for her freezer and washing machine. She lives her life outside as she has always done.

There were many painful moments as vandalism undid our work. Very often the kids destroy the trees, but it's a learning experience for both of us. One summer all the kids had toy spears cut from the leucaenas [*Leucaena leucocephala*] planted on the western and northern fence lines of the clinic. They were running around imitating Shaka Zulu, a television program about the Zulu wars. The trees were all snapped in half, but one must adopt a changed perception that this is an actual use. If the kids are needing to make toy spears out of the trees I plant, then there is surely a demand. In the perceived destruction there is an inherent use. It is just a matter of perspective because there are different cultural views. If all the trees that I have planted in Kintore had survived, the community would look very different. But many of the trees do recover as the bush recovered for thousands of years prior to contact. The leucaenas did regrow and now provide needed shade around the medical centre.

I now have a list of species for shelter belts that are domestically untrashable. All species have been selected on quick growth rates with the capacity to root sucker or germinate from seed in and around the planting. I recently cut out a 5-year-old *Acacia ligulata* [Sandhill Wattle] that had been eaten by termites, and underneath there were no less than fifteen seedlings, some already 30 centimetres high. Fire resistance is a selection criterion and so is the ability to cope with regular pruning for green manure or fresh organic matter. All plantings slowly mulch themselves

The council area in Kintore shows the impact of the planting efforts.

by catching windblown rubbish and I have resolved to use only weedings and low nitrogen organic matter as a top dressing to hide any possible eye-sore. It is imperative that this is held down with bits of scrap metal so the gales don't respread it around the community.

Ground covers are the solution to the dust. Everybody wants a lawn but has no idea of the maintenance or cost involved.[9] For five years I resisted having anything to do with lawns until I found lippia [*Lippia sp.*]. Now we have three lawns around the council area for people to sit on and these remain flat, unlike Couch Grass, which grows on and in the french drains becoming a thick belt of rubbish and rhizomes which block the septic flow within five years.

The increased vegetation also helps produce the biomass that can be used to attain a system of food. Every time I organise a new planting with an irrigation drip-line, in eight to twelve months it becomes a recreational program for every little hunter-gatherer. The kids diligently hunt every lizard and bird, pull out drippers for a drink and pick billy-cans full of solanums,[10] which are important sources of vitamin C. So there is a nutrition component to these plantings.

Also, welfare payments never seem to last the full fortnight and people very often spend three or four days living on damper, tea and sugar, just as stockmen did during the last depression. However, all the legume or seed-bearing acacias are represented in the community and anywhere you place irrigation for community plantings you get occurrences of important dietary grains.

I see that the future for improving Aboriginal health lies in environmental health and the production of food. Permaculture is the mechanism to do this and permaculture depends upon the local gene pools of indigenous plants in low maintenance areas to protect a more intensive food production area— after all you can't go wrong with two hundred million years of genetic information. There is no point in having flowering fruit trees with no shelter-belt protection when the next 70 kilometre-an-hour wind blast-strips the trees, preventing them from flowering.

'Mrs Bennett is one person who loves her trees as she loves her dogs.'

Prior to contact the Pintupi used vegetation as their principal resource and their game was specifically habitat-based. Managing each habitat was usually

*Bush Tomatoes (*Solanum chippendalei*) are an important vitamin C source and proliferate in the community.*

by fire; they burnt country and moved on without leaving any waste behind. Now they have sat down in one place but have not yet encompassed the change from hunter-gatherer to a horticulturally sensitive society, so fire can be a very destructive mechanism. Fire is still a tool and, although the herbs and grasses which grow around plantings are vital for stopping dust carried in the first metre from the ground, they are often burnt for fear of snakes or because flicking a match is an easy way of weeding.

Fire remains culturally important as everybody cooks and sleeps by a fire. Mulga [*Acacia aneura*] is the major wood used for cooking and weaponry while Gidgee [*Acacia pruinocarpa*] is a second choice.[11] Unlike the majority of central Australian Aboriginal communities, Kintore has a limited supply of Mulga woodland, which is greatly in need of regeneration. Other communities have stretches of Mulga up to 60 kilometres wide while Kintore has one area 7 kilometres by 2 kilometres and another 5 kilometres by 4 kilometres—that's it. It is some consolation that only dead trees are taken, usually the ones killed by fires or termites. However, judging from their size the trees selected are at least thirty years old. As the community has only been in existence for fifteen years and regenerating seedlings are driven over in daily collection attempts by vehicles, tractors and trucks which compact the fragile soils, the Mulga resource is unsustainable. The need for firewood is presently not apparent but thirty years in the future there will be a crisis.

Both Mulga and Gidgee are slow growing, so I never saw the answer to regeneration in planting them throughout the community. Using water harvesting techniques on the foothills and ponding banks on the creek wash-outs and clay/loam areas to maximise the productivity of irregular rains could be a possible solution to help regenerate seedlings. As an alternative I've been planting Redgums [*Eucalyptus camaldulensis*] because they grow 2 metres a year and can be coppiced after five years. One fallen tree that was planted in 1985 was cut up in 1993 and delivered to the old women for firewood, but it was promptly turned into artefacts.

If all watering was ceased at Kintore the first dead patch would be the vegetable garden. Worms

would have to run for cover and probably seek refuge in the french drain system. Organic matter would cease breaking down and the termites would advance at a great rate. The next set to go would be the fruit trees and any tropical trees that were not drought resistant. Tropical varieties of drought tolerance capacity would continue for many years as long as some water harvesting mechanism was in place to maximise rainwater catchment. Swaling or ponding banks with or without some mulching system would be an example of water harvesting mechanisms. Most indigenous trees would persist and grow after rain, set seed and continue as a changing plant community through drought and fire for successive generations.

It has never been easy at Kintore but it is happy work when you finally manage to beat the odds. Many people would have given it up as hopeless but I made an unwavering commitment on the basis that these people are committed to living here and not moving anywhere else. I respect people who have a long-held attachment to place as an identity. I've felt that I've had a home and I've never felt that before. I have a sense of belonging and a respect for country.

'I respect people who have a long-held attachment to place as an identity.'

Notes

[1] *Yanangu* is commonly used as the term for Pintupi people and is defined as 'people; animals; objects; used to contrast reality with unreal objects; a real body, not a spirit'. (Hansen and Hansen 1992)

[2] Of 72 people brought into Papunya in 1963–64, 29 were dead by August 1964 (Pintupi Homelands Health Service 1994).

[3] Outstations are small Aboriginal settlements or homeland centres, usually of people with close kinship ties, in lands significant to them for cultural and/or economic reasons.

[4] Barlow, C. (ed.) 1991, *From Gaia to Selfish Genes*, Massachusetts Institute of Technology, Cambridge MA. Nigel requested that this quote begin his story, stating that 'This is what I have done with rubbish'.

[5] *Albizia lebbek* (Siris Tree) and *Leucaena leucocephala* (Sneaky Tree).

[6] *Erythrina verspetilo* (Batswing Coral Tree), *Atalaya hemiglauca* (Western Whitewood) and *Eucalyptus camaldulensis* (Redgum).

[7] From P. Rivalland, *c.* 1988, headed 'Environmental change

and health in Kintore' and discussing the work of Kintore's Parks and Gardens team (in possession of RMIT OutReach Australia Program, Melbourne).

[8] *Wilytja* is usually defined as 'shade or sun shelter' (Hansen and Hansen 1992), but also refers to contemporary shelter structures (Sinatra & Murphy 1997).

[9] Central Australian lawns cost approximately $4 per square foot (about $14 per square metre) per year in watering.

[10] *Solanum centrale* (Bush Raisin), *Solanum chippendalei* (Bush Tomato) and *Solanum diversiflorum.*

[11] Mulga is an Aboriginal term for a long narrow shield made from acacia wood. Gidgee, also known as Black Gidgee and Tawu, is also the standard trade name for *Acacia cambagei*, which has a wide distribution in central-western Queensland and the Northern Territory, but not as far west as the Kintore area (Turnball 1986).

I love this land, I was born here

Bob Purvis

My father was a European and this is where he chose to live. Of all the things he understood and was good at, he didn't really understand this country. I'm born here and I love this land. I see things that he didn't see.

Out of the 800 square miles that make up 'Atartinga', there are 200 square miles that you can use for grazing. If there was not stock, it may revert to what it once was, but since the time the Europeans took over the land you have this massive movement of soil. See you've got an extremely fragile landscape that has never had cloven-footed or hard-footed animals on it.

I'm one of those people who is trying to hang on to what we've got. I'm trying to piece together what this land once was, and we've got a bit of a handle on it. We have become relatively successful at reclaiming land and the only way you can reclaim it is to build ponding banks. All you're doing is catching nutrient and moisture. I've put in about 450 ponding banks now over twenty-three years.

I see the destruction of this country and I dare say somehow that affects you deeply. I would like to leave the property to my children with the erosion in a relatively stable state and the land not degraded. I would like them to have the knowledge and the will to care for it.

'Atartinga Station'

The vast Australian arid and semi-arid zones support an extensive pastoral industry, which is responsible for managing the rangelands. A poor understanding of the management requirements for these sensitive areas since they were opened to grazing has resulted in overgrazing by both domestic and wild animals, and inappropriate fire regimes.[1] In addition, over two hundred years of tree clearing, the impact by hard-hoofed animals on the landscape, and the introduction of weed and pest animals have caused soils to erode, pastures to decline, and extinction of native plant and animal species.[2] Many pastoral areas consequently experience declining productivity, requiring management strategies that acknowledge the variability of the rangelands to maintain both productivity and bio-diversity, the importance of reducing grazing pressure, and the role of fire in rehabilitating land.

'Atartinga Station' lies about 200 kilometres north of Alice Springs in the heart of central Australia. The property of 2235 square kilometres consists of three main types of country—hard Mulga (*Acacia aneura*), spinifex (*Spinifex sp.*) sand plain and open woodland, and smaller areas of mixed sweet calcareous country. Robert Henry Purvis selected the land because of its shallow ground-water supplies, and sank the first bore in about 1918. He permanently settled the run in 1924, naming it 'Woodgreen'. Initially breeding sheep, and horses for the Indian Army remount trade, Robert Purvis turned to cattle in the 1930s. He used the good country for maximum grazing, but his lack of knowledge about arid zone pastures resulted in wholesale pasture and land degradation, and near-permanent drought.

Robert Purvis's son Bob developed an intimate understanding of the property whilst growing up on 'Woodgreen'. He acquired the property from his father in 1960 and started to implement management schemes to boost its environmental condition and economic production. His goals for improvement included reclaiming degraded land, developing a productive breeding herd, overcoming the property's debt, and developing an appropriate management philosophy.

Bob began to reduce the number of cattle on the property to decrease their impact on the landscape. Establishing the most appropriate stocking rate for the property is a continuing process, and it took some twenty years for Bob to feel that he was getting close to understanding how much stock his land can carry. To reduce his stocking rate and maintain economic viability Bob had to improve his returns from each animal. He developed a breeding and culling program over fifteen years, using Poll Shorthorn bulls to raise an even-tempered animal more adapted to the climate and pastures. The heavier weight and superior quality of the animals have increased his returns, despite a reduction of his herd from its pre-1970 figure of 3000 to 900 in 1995. The herd size is maintained through sales of bullocks, aged cows and surplus heifers.

When Bob took over the property, he also inherited a debt in excess of the property's value. To pay off the debt and support his family he worked elsewhere for wages as well as on the property. In 1971 he bought 470 square kilometres of the neighbouring 'Waite River' run on the eastern boundary of 'Woodgreen', renaming the enlarged property 'Atartinga'.[3] The purchase increased the amount of good country from 130 to 450 square kilometres, enabling the Purvises to better service the debt and buy equipment for land improvement works.

Bob recognised that time alone would not repair the damage of past grazing practices, as some areas had been destocked for twenty-five years with no improved land health. As most of the nutrients in the soil are in the top layer, Bob understood the importance of retaining water on potentially productive land rather than letting it flow into the creek systems, taking the nutrients with it to feed thick, woody weed-infested areas. He decided that an essential part of his approach to reclaiming the land would be through water harvesting: developing ponding banks to capture water and allowing it to stand long enough to infiltrate the ground.[4] Trial and error guided Bob to better understand and refine the ponding banks that he was building—developing a balance between the area and depth of water being retained and the runoff characteristics of soil types. Over 600 banks are now part of the 'Atartinga Station' landscape.

Another vital component of Bob's management strategy on 'Atartinga' has been pasture improvement. It was necessary to introduce a hardy perennial to replace many of the fragile native perennials eliminated due to overgrazing. Buffel grass (*Cenchrus ciliaris*), originating from Africa, was chosen for its stock palatability and quick soil cover response. The use of fire as a regenerative tool opened the way to increased grass growth where degradation had led to dense shrub encroachment in water courses and open woodland country.

In addition to burning practices, animal and land stress have been reduced by integrating the herd with the landscape. The poorest country is not grazed, while the good country is grazed by marketable stock. Adjoining paddocks are managed in pairs with cattle spending six months in each paddock.[5] At least two or three paddocks remain ungrazed at one time, so each paddock rests for twelve months over a five-year cycle. The breeding herd runs in small groups of 50–80 cows in a paddock, with the bulls remaining with the cows. Water on the property is kept clean by covering the tanks, which also reduces evaporation, and is piped to the preferred grazing areas to help improve the quality of the cattle's nutritional intake and decrease grazing pressure on the pasture.

Bob Purvis has devoted nearly thirty years to improving the health of his land. He notes that to this day many central Australian farmers retain the government-endorsed grazing philosophy that nearly ruined 'Atartinga'.[6] The success of Bob's management strategies was indicated in 1985 when the last of the debt on the property was paid, and the heaviest bullocks were sold in the drier-than-average season while the rest of the pastoral industry in the Alice Springs region was in crisis (Purvis 1986).

Atartinga homestead

Bob Purvis, a grazier in the arid interior of Australia, is leading the way in creating sustainable farming in the Northern Territory. He stands on a slight rise in the landscape on his property, 'Atartinga', and points to the expanse of his most degraded land: 'This is our worst piece of land and much of central Australia is like this'.

This scene is testament to the condition of large tracts of land declining under the pressure of past and current grazing practices, not only in arid regions but across extensive areas of agricultural lands throughout the country. Bob has devoted nearly thirty years to improving the property through reducing the stocking rate, implementing a breeding program for his cattle, establishing a stocking cycle for the paddocks, developing a water-harvesting program through building over 600 ponding banks, improving pastures and implementing burning practices. In doing so, Bob has developed a more sensitive system of grazing the property with reduced pressure on the environment and its resources, and greater economic returns.

Bob

My father was a European and this is where he chose to live. He was a man of many parts. Of all the things he understood and was good at, he didn't really understand this country. No better or worse than others of his generation.

I'm born here and I love this land. You take on board a lot of things that a blackfella would, without even realising you arc like that. I see things that he didn't see, but you don't realise that that's how you are.

Our soils are only two-thirds as good as other deserts in the world. Half of the nutrients in this soil are in the top half centimetre—half! If you lose that top centimetre, you're down the drain. The soil is so poor that nothing can live on it. Well, a few kangaroos, termites and lizards, but not stock. Out of the 800 square miles [about 2000 square kilometres] that make up 'Atartinga' there are 200 square miles [about 500 square kilometres] that you can use for grazing, and only some of it is in one lump. With fencing, you can have your stock using a bit less than half of that useable country, so at the end of six months you shift the stock on to the country that is not being used. But those pieces of good country are small, and this contributed to what went wrong here, why my father went broke. He didn't think there was much difference between the good country and the bad, and yet it's the difference between chalk and cheese.[7]

Our soils are only two-thirds as good as other deserts in the world. We can turn that around and say they are by far the poorest soil on earth—I mean nutrient-wise.

When my father stocked this country the cattle could eat about 5 per cent of the plants, and they knocked that 5 per cent out straight away. Let's say it took five years to eliminate those plants from the landscape, because most of the damage here is done within five years of first being stocked. And 75 per cent of the property is like that. You can't really use it. So he was back to dying cattle all the time. Cattle died here every year, all year.

Nearly all of our grasses were made for kangaroos and birds; they were not made for cattle. They are very fragile and the cattle ate them—knocked

Gully erosion in a fertile watercourse, 1967

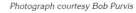
Photograph courtesy Bob Purvis

them out. It is very difficult to reintroduce indigenous grasses because you don't know where they were, and the good Lord had them so they only grew in small areas. If you happen to find a small area where they grow, you're right, you'll succeed. But if you plant in an area where they didn't belong, you will have total failure all the time. But Buffel, the American grass that originates from southern Africa, and some other exotic grasses, will grow over a wide spectrum of soils and you have greater success. Our best perennials are very good, very nutritious, but they can only stand the right grazing pressure, compared to Buffel which is the most successful thing we've introduced.

The condition of the land may be due to a very big natural cycle, but there's a new element in the ball game because we've introduced stock. If there was not stock, it may revert to what it once was, but since the time the Europeans took over the land you have this massive movement of soil—you've lost the best of the soil. See, you've got an extremely fragile landscape that has never had cloven-footed or hard-footed animals on it. It's got a variety of plant species that have never grown up with that animal. You put cattle on it; it's like introducing smallpox into a race of people who have no immunity to smallpox. Well, that's basically what happened to this landscape here.

The piece of country that my father originally took over is leased land, it's not freehold. The lease states that you have a minimum stocking rate of 3000 head for the property. We reckon that a sustainable number is about 400. You can't make a living on 400 head of cattle, so we were fortunate enough to buy more land. But the official stocking rate on that increased amount of land was 4500, the minimum stocking rate directed by the government [Northern Territory Department of Lands]. We run 1100 [in 1992]. That may or may not be sustainable—it quite possibly is.

In good heart, this country can only support a stocking rate of about 7 or 8 beasts per square mile [about 3 per square kilometre]. That's what the land can carry, and my father had to stock 60 to the square mile [about 23 per square kilometre]. He had 50 square miles of useable country in amongst 500 square miles and a minimum stocking rate of 3000.

Photograph courtesy Bob Purvis

*Eroded landscape north of
Acoota Boundary, 1968:
'I see the destruction of this
country and I dare say somehow
that affects you deeply.'*

North of Acoota Boundary, 1984

Photographs courtesy Bob Purvis

North of Acoota Boundary, 1990

The cattle won't live in that poor country; they have to live on that 50 square miles, so you're looking at around 60 per square mile. That is why it went to pieces. There is no mystery about that. I don't see that you can manage the land any other way but to reduce your number of stock if you want to hang on to the land. Once you eat out the grass, it won't grow on the next rain or maybe two years' rain. It takes years. Sometimes it takes ten or twelve years for grass to grow again once you've removed it totally. The nutrient level of the soil is so low that you can't just eat it all off and expect that it'll grow again.

So for a bureaucrat to put that stocking rate of 3000 down and then for my father to go broke, I know the stocking rate is far too high—it's obvious! So I reduced it, even though we were in default. See, once we get below that rate of 3000 we're in default of government policy, see we're in default all the time. You've got to convince the bureaucrat that the stocking rate is unrealistic. It took me twenty-odd years to get them to change, and it only changed for me. Although it did show that there was something seriously wrong with stocking rates in this area it is difficult to convince that those minimum stocking rates should be maximum stocking rates because the politician, he wants the rent now. He wants it yesterday, and your rent is based on the property's carrying capacity.

The land is what limits you. We have to be able to produce good cattle on degraded land and improve that land at the same time. Once you have rehabilitated all that you can win back, twenty or thirty years down the track you may be able to carry a few more cattle. We are only limited by that basic resource. I'm one of those people who is trying to hang on to what we've got. The Australian attitude—or this government's attitude—is to plunder that resource, you know, the finite resource, not the infinite resource.

I suppose when I first thought about this I said to others, 'We ought to be managing this country better'. But they just said, 'You're broke and nobody will take any notice of you!' It's a fair comment, but it took me another twenty-five years to prove those ideas. And now a lot of these fellas are broke and we are not. They reckon that the seasons have changed, or the rain has stopped, or it's an economic downturn.

Their land is going backwards, but they don't want to know and they don't want to see.

I had a crumbling property to take on. To hang on to that I was going to have to do two men's work. I was going to have to work in the vicinity of eighty to ninety hours a week. You don't have any time to go anywhere else to learn, and there is no money anyway. You're on your own for years because you have to teach yourself, you see. I went to school, sure, but none of the things I wanted to learn could I learn at school. All of the things that were rammed down my neck were useless to me and, unfortunately, I realised that when I was a kid. You know, if you were ignorant of that you'd be right. So the hardest part was to obtain the expertise early on, to teach myself, and because all the landscapes are slightly different, what applies here doesn't necessarily apply somewhere else, so I don't know until I've wrestled with it.

I am trying to piece together what this land once was, and we've got a bit of a handle on it. And, of course, when you start on one of these things, you've got your ideas, but as you go along you realise you have to modify and you change as you learn all the time. You still don't know whether you're right or not. Each year is a little bit different, but we have become relatively successful at reclaiming land and the only way you can reclaim it is to build ponding banks. The principle is simple. The Incas woke up to it and

'Time alone would not repair the damage of past grazing practices.'

the Romans woke up to it. All you're doing is catching nutrient and moisture.

Ponding banks collect silt and have to be able to withstand torrential rains in the order of one 100 millimetres in two hours. Our average annual rainfall is 250 millimetres—but it is erratic. The nutrients are in the silt. If you have bare, eroded country, each time one of these flows occur the silt will all go down the creeks.

We wrestled with how much you can pond and not saturate that ground and kill all the bugs that are in it, all the plants. If you have very hard ground, 9 inches [nearly 23 centimetres] of water will lay for eight or maybe ten days when you first build the bank. Initially a few plants will come on with that first rain, and they will be worthless plants [for grazing] but they start to penetrate that ground and that soil modifies. Three or five or eight years later, you will have gone from those undesirable plants to desirable plants. As it becomes modified the ground is building nutrients and eventually it becomes a rich area. By trial and error, I worked out that 1 metre is an optimum height for the bank to hold 9 inches of water. If it was a foot [30 centimetres] of water, it would stand a little too long, the soil becomes water-logged and plants die.

Umberumba, 1990; ponding bank built in 1988

When you are trying to turn this around you have enormous runoff, but gradually, as you get more

Photograph courtesy Bob Purvis

vegetation on the ground, the flows will slow. When we kicked off, if you had a big rain here, everything would have stopped running within an hour or hour and a half. Now, twenty to thirty hours after rain, the banks will still be spilling water. The water is running so much slower, higher up. The silt is caught in the pond and clean water is spilled into the creek. The banks vary from 50 metres to 250 metres in length. With age they become part of the landscape. You can't even see that they are there. There are some ponding banks totally covered with grass and you wouldn't know they were there until you fell over them —that's your ideal.

'Sometimes the landscape allows you to add something to it, and it just fits in.'

I've put in about 450 ponding banks [by 1992] now over twenty-odd years. We have never had any funding. I'm assuming you have to do it yourself. But you must realise that I had to teach myself how to do this. I observed what happened to the soil during big rains, and as I lay thinking about the problem, build- ing ponding banks seemed to be the only answer for stopping this erosion. But who do you go to for help? You'll go to various people and find that everybody likes to give you advice, but you find out after four or five years of work that you might have been better off never going near them.

So, 450 ponding banks in twenty-odd years doesn't seem many, but five or six or eight years of that was learning how to do it. Whenever you are teaching yourself it takes a long time. If somebody had the knowledge they could maybe teach you in one week that which took me five years to learn. I tried many avenues until I finally started to go the right way and other farmers in the region have followed the example.

It could be argued that this land should never have been stocked, but I didn't have that choice. We now know that if you wish to stock it, or if you wish to have all your birds and all your other things on it, you've got to get it as close to what it was from the fire regime standpoint. What kept the status quo in this country was the fact that you had a certain amount of vegetation and you had wildfires. We don't know how often. I mean, it varied on the vegetation and the areas; sometimes it may have only been a fire in fifty years, sometimes it might have been a fire every fifteen years. We don't know. But what we do know was that there were fires. Now, the bureaucrat comes here and he eliminates fire, so this country here probably hasn't had a fire for eighty years. You've eaten it bare, so nature grows trees to protect the land. Instead of a grassland with a few trees in it, you have dense scrub—6000 trees per hectare, and that's useless to anybody.[8] It's completely worthless.

You can't ever have it like it was, but you realise all things are intertwined in that ecology, and to try and get some balance back, fire is one tool you must have. Fire was not a human-made thing. It was a part of the landscape, and if you try and remove it totally, you produce a totally different landscape. The blackfella burned, but my father's generation stopped the blackfella from burning and the blackfella has another generation that is not taught how to burn. And then there's my generation, the forty to sixty year age group, who comes along and has to learn what the blackfellas knew three generations back. See, there's nobody left. The blackfella today doesn't know—I got to teach the blackfella. And you've got to try and teach the pastoralist and the greenie, because he hates fire, because fire is seen to be so destructive. You can't remove fire: fire is part of this ecology.

We are getting better at it. We are getting closer to the original fire regime, but I wouldn't know how close we are. So you have to have fire, and you have to have enough fuel, vegetation litter on that ground, to be able to carry a fire. And one fire is not enough. If you haven't had a fire for eighty years, it is more important what happens after your first fire than before. I mean, sometimes things go against you. If you had, say, one fire and it didn't burn fiercely enough, it removes all your fuel without burning your trees. Well, then you are still at square one. We can't predict what's going to happen afterwards because there's a wide variety of things that can happen, but if you can get a second fire within five years, you're just about on top of it. In some country we have had three fires now within twenty years and I think that is almost what it was when my father came here. It's one of the tools you have to have.

I see the destruction of this country and I dare say somehow that affects you deeply. That is worth more than anything else. It's worth more than money. Somehow it's a very valuable thing. It's the thing that drives you. And if you're completely broke, what do you do about it? The last thing you are going to do is walk off. You just have to get better at looking after the land. That's why we are where we are. And I'm not saying that another person might not do a lot better than I am, or what I have done.

We have dropped our stocking rate considerably and we make more money. So we said, well, how far can you drop that stocking rate and still maintain that same money or more? We reduced our cow numbers from 1000 to 400 to 300, with almost the same money, giving our land a much better chance. All you have to do is keep that stocking rate low enough and the land is healing itself. It may be another forty or fifty years before the plants that were once here will grow again. But because it's not rapidly eroding I don't have to do anything to it except see that not too many stock graze on it and the soil doesn't wash away.

'The condition of the land may be due to a very big natural cycle, but there's a new element in the ball game because we've introduced stock.'

See, an average grazier comes here and straight away he's working out how many more cattle he can run. In the first year or two he would make a lot more money than me, but then he'll wonder what the hell ever happened to him and soon it will be gone. When it goes, all his birds will disappear and everything else. All you've got left is a few crows and a few eagles, anything that will live on a dying landscape.

We would not be here if we hadn't managed this land better, and that is all I have tried to do. As we have managed it better, so we have learned how to manage it better again, and no doubt somebody will come along and do a much better job than I have. But you must arrest the decline. I see it as my job to stop that slide. Because that is all you are going to have time to do in one lifetime. I would like to leave the property to my children with the erosion in a relatively stable state and the land not degraded. I would like them to have the knowledge and the will to care for it.

I love this land, I was born here. Why would I want to live anywhere else? I will cease work when my body can't work any more.

From a letter from Bob Purvis, 24 December 1994

Time and change go together. The drought is taking its course, our driest year since 1919 was 1961 when 40 millimetres of rain fell. This year is similar. We had one rain gauge then and nine now so we have a better idea of storm activity. Rain in the nine gauges varies from 16 millimetres to 62 millimetres. We have lost no cattle this year, in 1961 half the herd died.

The number of ponding banks has progressed to six hundred.

Notes

[1] Rangelands are areas where rainfall is too low or unpredictable to sustain intensive agriculture, and soils are highly weathered and infertile. In Australia they make up about 70 per cent of the continent's land mass, of which large areas have suffered some degree of degradation. Fire was an important land and economic management tool used by Aboriginal people. When grass was long and ground litter was thick, the country was 'cleaned' by setting fires, which promoted the growth of food plants that would otherwise be choked out, created favourable conditions for kangaroos, and reduced the risk of more serious fires. The subsequent mosaic pattern of the landscape represented various degrees of recovery after fire (CSIRO 1991a, 1991b).

[2] Only 10 per cent of shrublands and 30 per cent of the grasslands approximate their original state; 65 per cent of the shrublands and 15 per cent of the grasslands have degenerated to less than 60 per cent of their pristine condition (Hall *et al.* 1972).

[3] *Atartinga* is an Aboriginal term for the ironwoods (*Acacia estrophiolata*) that grow in the homestead area.

[4] The value of this work was shown three years after commencement when, from an area reclaimed with ponding banks, fat cattle were turned out for the first time in twenty years (Purvis 1986).

[5] Several sites show an increase of these palatable pasture species: Curly Windmill Grass (*Enteropogon acicularis*) in gidyea woodland and Woollyoat Grass (*Enneapogon polyphyllus*) in open woodland.

[6] CSIRO emphasises the need for more appropriate land management philosophies to ensure that the pastoral industry in the arid region survives: 'Better herd management can improve animal quality and reproductive efficiency. By producing fewer animals for specialist high return markets, pastoralists can stabilise their incomes and reduce risk. Pastoral enterprises which rely on high stock numbers as insurance against fluctuating markets and seasons run the risk of widespread pasture degradation and reduced carrying capacity' (CSIRO 1991b).

[7] In a landscape that may appear flat and featureless, there is a diversity in productivity. Water and nutrients concentrate in patches, around vegetation, or in gentle depressions, creating sites that are the key to the productivity of the area (CSIRO 1991a).

[8] Due to pastoralists not actively burning, many of the emerging inedible trees and shrubs that germinate after rains become established, reducing the amount of edible grass available for grazing (CSIRO 1991a).

I never ever bought water yet

Karl Hammermeiste

I've been twenty-two years in Coober Pedy. For seventeen years I never go without water, and I never bought water yet. I catch that water and it runs into the underground tanks. Every tank is full. Now, if I found a little bit of opal, I would put another one there because you get those downpours, you get all the water you want in ten to fifteen minutes.

When we came to Coober Pedy twenty odd years ago, we had no tree and not a bush—absolutely nothing! But look at the gums, how they grow. They grow so beautiful. They are only five years old.

The main reason for doing all this—trees and water—is the environment. My neighbours, they plant a lot of trees lately and it's good. The town looks much better I think, and it's good for the people, but there aren't too many people in town that are able to get through the whole year with their own water supply.

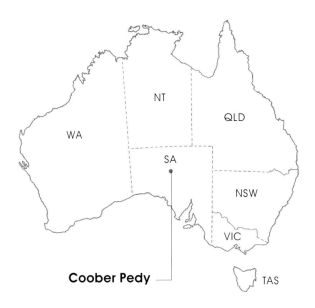

Coober Pedy

Coober Pedy[1] lies approximately 750 kilometres north of Adelaide in the expanse of ancient sandstone and claystone of the Stony Desert in South Australia. The region has an erratic rainfall with an annual minimum of approximately 125 millimetres. The impact of low rainfall is exaggerated by summer temperatures reaching 50 degrees and an extremely high evaporation rate. Local vegetation is very sparse, consisting of low chenopod shrubland[2] with ephemerals after rain, and small trees and shrubs along the creek lines.

Between 15 and 30 million years ago silica and water were trapped beneath subsurface rock in an escarpment in the Stuart Ranges, creating the fire in the opal stone that continues to attract miners and buyers from around the world. This water gem drew people to live in a landscape seemingly devoid of any water.

Opal was first discovered in 1915 by a 14-year-old boy who was camping in the area with his father's gold prospecting party. A considerable mining field was established by 1916 and Coober Pedy township began to develop. Major finds stimulated mining in 1946, and the South Australian fields now produce 80 per cent of the world's opal.

Until recently, the Coober Pedy fields did not have a reliable supply of adequate water. In the early days, miners used 20 gallon drums to store grey water, waiting for the sediment to settle before boiling the water twice for washing, and returning it to the drum (Crilly 1990). The drums were rationed at one per person per fortnight. Water was expensive,

Coober Pedy's reverse osmosis plant, 23 kilometres north of town

costing £3 for 100 gallons when available, and having to be carted from twenty to forty miles away as the closer waterholes dried (Crilly 1990). Bores were sunk after 1917, but the water had a high salt content. The South Australian government built an underground storage tank with a 2.27 million litre capacity in 1921. In 1966 solar stills were built to desalinate the water, later replaced by a reverse osmosis plant, then a new plant in 1979. In 1976 a survey located suitable groundwater 23 kilometres north of Coober Pedy, resulting in a new bore and water with a much lower salt content. A 1981 feasibility report for supplying reticulated water from the bore to the town resulted in a new reverse osmosis plant being commissioned in 1985 to supply high-grade potable reticulated water. (Layson n.d.). However, water remains an expensive resource at $5 per kilolitre (in 1992) plus a $20 fee for delivery if it is carted and not supplied through reticulation.[3]

One of Coober Pedy's opal fields

KARL Hammermeiste lives with his family in a dug-out in Coober Pedy.[4] Karl recognises that, in a landscape where the creek lines flow with little more than dry river stones, when the rains do arrive they provide an ample quantity of water to harvest. He has created a garden in this arid environment by designing the verandah to his dugout as a water catchment, from which rain-water is piped to underground storage tanks with a collective capacity of 200 000 litres. Despite the limited and erratic rainfall, Karl recognises that the real issue is not the amount of water available but the capacity to catch and store it quickly. Most of the population relies on a town bore, but Karl's recognition of his environment has allowed him to live in Coober Pedy for seventeen years without buying water. It has also allowed him to establish a garden whose green and sheltered environment contrasts with his dusty working conditions underground.

Karl

A water catchment verandah with pipes connecting to underground storage tanks shades the front of Karl's dugout.

I come 1961 to Australia from Germany; I've been twenty-two years in Coober Pedy, mining. For seventeen years I never go without water, and I never bought water yet. Never ever! Not one single litre, and there's three kids—the family. I mean we don't waste—we don't stand too long under the shower and keep

wasting water. Underground tanks, that's how I catch my water. You see, the roof here is 35 feet wide [about 10.5 metres] and that concrete on top of it makes about 41 or 42 feet by 174 feet in length [about 12.5 by 53 metres]. That's a fair bit of catchment. Its amazing how much water you can catch when you have that. I catch that water and it runs into the underground tanks. They hold roughly 200 000 litres. If we could store the water from our dam, which holds approximately 3 to 4 million litres if full, that would be a big bonus. I scoop the tank every year. I scooped it out once or twice and it's about 50 or 60 feet deep [15 to 18 metres]. Every tank is full. So if it doesn't rain any more until next year summer, it doesn't matter to me—or another year and a half, I don't care. I have enough water.

Years ago I had six 2000 gallon tanks. That is just over 7500 litres. I had over 45 000 litres in tanks standing. Up top, that's a 27 000 litre tank. Now, if I

found a little bit of opal, I would put another one there because you get those downpours, you get all the water you want in ten to fifteen minutes, as long as you can handle it and store it. And it stays fresh, cold —very cold! Oh, beautiful in summer time when you pump it out. Nice when you shower. Very soft, the water—beautiful.

We catch water and I soak the gum trees with the water from the pump, or when I'm not home and all my underground tanks are full, the last tank overflows. That's only for the trees. It soaks right into the ground. The gum trees are growing very well all by themselves, you know. They're not doing bad when you really think, for only that little bit water, and that excess water what comes out of the showers and things like that. That's not bad.

When we came to Coober Pedy twenty odd years ago, we had no tree and not a bush—absolute nothing! Can you imagine the place totally without

Karl's son opens the lid to one of the water tanks.

trees? Not a single tree when we came here; there
was not a single plant. I think in the start we planted
the wrong ones; we never knew the right tree up here.
The government gave the Peppercorn tree [*Schinus
sp.*] and they were all the wrong trees for here. I've
pulled out I don't know how many pepper trees. You
have to supply them constantly with water. And there
was still white donkeys around and they ate the lot.
They came in here and they would take that bush and
eat it to bits and pieces. Wherever a little bit green,
they chewed around on it. Yeah, we just listen to the
fellas, whatever they suggesting to plant. But look at
them gums, how they grow. They grow so beautiful.
They are only five years old.

The main reason for doing all this—trees and
water—is the environment. You know, when you work
every day underground in dirt and dust, you come
home and you sit again in dust. We get that thick
gravel and spread that around, otherwise you have
dust everywhere. And the trees help. The tree makes
oxygen - it's good for everyone. I think like that, you
know? My neighbours, they plant a lot of trees lately

and it's good. The town looks much better I think, and it's good for the people, but there aren't too many people in town that are able to get through the whole year with their own water supply—I don't think anybody. Most people here, they would pay 60, 70, 80 dollars a month for reticulated water from the town bore. Why should they? I find that stupid. It's up to the individual what not to do.

I will stay in Coober Pedy. I tried a few times [to leave]. I went to Europe—had a look. I go to Adelaide or somewhere, and after two days I can't stand it any more. It makes me sick. People run around. Individual people, they don't fit into cities and towns; they have to go out in the bush and do what they want to do. Good people—a bit rough sometimes, but that's how it goes. And look at my kids—they keep the cars, the trucks, the drills, the blowers, the machines—they tried it all. And they can weld anything—repair all our tractors and engines. My kids do everything and they never learnt anything [in school].

This environment, you know, it's fantastic!

Overlooking Karl's garden from above his dugout's catchment roof

Notes

[1] The name originates from local Aboriginal words *kupa* (white man or uninitiated man) and *piti* (waterhole), meaning 'white man's burrow' (South Australian Department of Mining and Energy 1986).

[2] Low shrublands are dominated by the family Chenopodiaceae, especially the genera *Atriplex, Bassia,* and *Maireana* (Turnball 1986).

[3] In 1992 the same quantity of water cost 82 cents in Adelaide.

[4] Dugouts are excavated buildings that were introduced by ex-soldiers returned from World War I. They are a unique part of the Coober Pedy townscape and their continued use is attributed to them being cool during summer and warm during winter.

I miss the sight of water

Cliff Ferrall

*I packed everything up, all my personal belongings,
packed up what there was of it and came down to Coober
Pedy from Alice Springs. I had a lot of poking around to do
in this place first. That took a couple of years, but from
then on I got stuck into the garden. I couldn't live in this
place and have it absolutely bare. So with a little bit of
pride, I think I've improved things.*

*Water and soil are the biggest problems. I would spend a
bit more than half my water bill on the garden. If the water
was cheaper, 50 per cent cheaper, most of the people
would have gardens—they'd bung in trees. And there's
nothing that draws rain more than trees. Year after year the
clouds split because of the trees along the creek line. You
can see rain pouring down on the horizon.*

*The main reason for putting the four tanks around the
house was to get what water I could for free. The garden is
75 per cent of my interest in life.*

CLIFF Ferrall is retired and lives in Coober Pedy. He began developing his garden despite limited water and soil resources, and the hardships of maintaining it. Although his house is connected to the town water supply Cliff invested in two extra rain-water tanks due to the expense of reticulated water, increasing his water storage capacity to 31 000 litres and enabling him to maintain a garden with only a pension for income. Cliff identifies his interaction with a garden as a vital part of his being. In Coober Pedy's arid townscape, a garden not only provides shade and helps to reduce the movement of dust, it also provides company. Cliff spends a good portion of the day in his garden.

Cliff

I was in Alice Springs for fourteen years before I moved to Coober Pedy. Oh, I miss water. I miss the sight of water. I don't mean rushing down the beach and swimming and all that sort of thing, but the actual sight of it—the coolness—the sight of water. About twice a year the Todd River would run. You saw water there and it would lay there for a while—quite a while— and then it would dry out. You'd see no more water until the next rain.

I didn't intend coming to Coober Pedy at all, but when Alice Springs got too civilised, with their damned malls and darn traffic lights and junk, that was enough for me. I packed up to go to Western Australia. It was the only state in Australia that I hadn't been to. I've been to every other state. I've even been up on Thursday Island. But, without being morbid, I thought before I die I'd love to go over and see Western Australia. I'll go up and have a look at that northwest coast, along Port Hedland and Broome, places like that, and go down to the southwest corner and have a look at all the big timber down there. So I packed everything up, all my personal belongings, packed up what there was of it, and came down to Coober Pedy from Alice Springs.

I went to buy some stores then I went up the pub for a bottle of beer. They asked me if I wanted a job. I thought, well that'd be handy—a few bob [shillings] for petrol. So I stayed there for a little while and finished up here still! I got here and just settled down. If I pack up and move now I've only got to turn around and start making new friends, and I'm too old at eighty-three.

I had a lot of poking around to do in this place first. I had to build myself a lot of furniture because there was nothing in the town you could buy. I had to make the dresser, I had to make the table the sewing machine is on, and I had to make the writing desk. I had to make these damned things up myself - I made a bookcase. The verandah was an open verandah, so I closed it in and then bought the aluminium sliding windows. That took a couple of years, but from then on I got stuck into the garden. The garden and the chooks are my most important things to look after.

Wherever I've been I've had a garden. Up in Alice Springs I had a lovely garden - great big area, and chooks. I had two places up in Alice Springs. Tennant Creek was the same - in goes the garden and trees. I was in Tennant Creek for twenty-five years.

The garden is a hobby and it freshens things up a bit to see the greenery. There's a freshness to

*'I couldn't live in this house and
have that yard absolutely bare, it
would drive me mad!'*

having greenery around the place, and that's the main reason why I thought I'd have trees. But once I've stuck two or three decent-sized trees in the yard I wouldn't be able to shift my car around, so I've got to restrict it. I'd like to have a bigger place. If I had a bigger place I'd have more trees in the yard—nice big shady trees.

I leave a bit of me behind everywhere I move, in the way of greenery. I like to see a bit of greenery around the place, my word. You regret having to move. Once you've gone to the trouble of having everything looking nice and green, you're tearing a bit of yourself away. That, I think, is the idea of having a garden, or trees or anything green—to make a form of permanency, to make you feel as though you'd like to settle down for good. I couldn't live in this house and have that yard absolutely bare—it would drive me mad! When I first came here, that's what it was like—absolutely bare—bare as a badger's bum, the whole place! There was no little short fence, no greenery, no chook yard, no trees, no scrub even. So, with a little bit of pride, I think I've improved things.

That huge, big pepper tree [*Schinus sp*] out in the chook yard—everything in this garden in the way of trees and things, I bought just as small seedlings. Put them in the garden, watered them as much as I could—as much as I could afford to—and fortunately they were able to live.

Water and soil are the biggest problems with keeping a garden here. It's such a gravelly, soaking ground that it takes the water up.[1] You just can't splash it on and water the next plant. You've got to wait for it to get nice and wet, let it get a good soaking. No matter how good the soil is and how much mulch you put in with it, there's quite a bit of water that gets down away from the roots and you lose it.

I would spend a bit more than half my water bill on the garden. It would definitely be more than 50 per cent a year—that's 110 000 litres of water. I would never use 55 000 litres of water in the house—even showers and washing. For June, I would say approximately 3000 litres went on the garden, and the other 1300 litres would be in the house.

That's the nub of the whole thing in this town—the water. I keep pressing on it too. I've been here long enough to know that it is the main whinge in this town. It mightn't so much be the water—it's the price of it. When you consider $5 for 1000 litres of water up here and 82 cents for 1000 down in Adelaide, that's one hell of a difference, even allowing for it being isolated up here, and it's got to be pumped and put through the reverse osmosis system to get the clean, pure water for drinking. Even allowing for that, it should never be $4.20 dearer up here. And that is the big bugbear. If the water was a lot cheaper, 50 per cent cheaper, most of the people would have gardens —they'd bung in trees and things like that.

And there's nothing that draws rain more than trees. That's something else you learn if you are here long enough. I've been here thirteen years now, and every time it's going to rain, huge big clouds come over from the northwest—black as the ace of spades. They come right up and, without a word of a lie, they split! And as soon as they split, one lot will follow a line of trees, a whole heap of trees out southwest, and they'll follow them along there. The other lot will go out towards Oodnadatta. Year after year after year, the clouds split because of the trees along the creek line. And the next thing, you hear Oodnadatta have had a good fall of rain and that's only about 250 kilometres away from here. People say, 'Oh, the rain must be coming this time—black, beautiful clouds!' We've had storms with thunder and lightning, and they still split, and the clouds go right along the creek. You can see the rain pouring down on the horizon, and you say, 'Oodnadatta or Glendambo's getting that'.

'Water and soil are the biggest problems with keeping a garden here.'

We haven't had any rain for four years. Goddammit, there's children born in the town that haven't seen it yet! You have a bit of a scud across every now and again, so you walk outside and kick the ground, but it's still dry. When I'm talking about rain, I mean *rain*, when the water is running on the streets. About four or five years ago we had decent rain—beautiful! I had my tanks all filled up. Then you wait for the next lot, and the next lot didn't come until about '89 or so—we got three inches [about 76 millimetres]. That topped the tanks up a bit, but we haven't had anything in the way of rain since then.

The main reasons for putting the four tanks around the house was to get what water I could for free, and good water. That was the main reason I went to the extra expense of putting the tanks there. You'd be surprised how much rain I get off my workshop. Now I've got my chook house guttered too, and I have not put any water in my chook house tank for ages because there's always enough sprinkle of rain to keep that tank full. It's surprising how little an area can still catch the rain.

I knew that every so often you get a good rain—a deluge. So I thought to myself, well, even if it only rains a little, I'm still getting a couple of tanks filled. If I didn't have the extra tanks and it rains heavy, I'm going to curse to high heaven. So I had two more tanks built—two 2000 gallon tanks, and the town said, 'Cripes, what do you want with all those tanks?'

'It'll rain one day', I said. 'I'll have water, and you blokes will still be paying for it.'

'So I had two more tanks built.'

Water was $67 for 1000 gallons [about 4500 litres] in those days, but they still said, 'Oh no, that's only a waste of damned money!' I was the bloke who finished up in front, and I still do. It's surprising how many people have got so few tanks—surprising! I've always got water, even with those three inches a couple of years ago. My four tanks hold a total of 31 000 litres—that's a fair swag of water, and it's good water.

The garden is 75 per cent of my interest in life—I can only put it that way. I don't know about it making me any different mentally or anything like that, but if there were mobs of stuff growing here then I'd probably feel better for it. I'm always striving to try and improve it. That's why I spend so much time out there. If it wasn't for the garden, I'd be niggly because I wouldn't have anything to interest me. When I used to visit Adelaide I was missing my garden all the time. Gardening's definitely been a hobby of mine for as long as I can remember.

You want to know something? I talk to my garden. Now laugh, but there you are. That's how interested I am in it. Just the same as you would with your chooks—you walk in the yard and have a bit of a talk to them. If people hear me they think a man is not right in the head. Still, I'm enjoying myself. I don't expect anything to happen to advantage the plants or anything like that, but being on my own I feel that

they are sort of companions. You turn around and you just have a little talk to them—I'll have to talk a bit more to those tomatoes. Yes, it's definitely part and parcel of my being, the garden.

From letters from Cliff Ferrall, 22 March 1992

. . . Am squeezing in two or three more trees into my garden this winter—two Peppercorns and an Ellendale Mandarin.

Am trying to get some 'Old Man Saltbush' (grows to about six feet high and has a larger leaf than the usual type) but cannot get any seeds or seedlings anywhere . . .

We are still looking for a good rain. Heavy thunderstorms moving all around us. You may have heard of them on the radio rolling in from the west into our state with general rains predicted all over the state. (They forgot to say 'Except for the Coober Pedy area.')

'In my opinion, the more greenery you have the cooler things are. Definitely draws the coolness. You only got to get along that dry old creek, and get underneath some of those big old gums along there, and notice the difference.'

21 February 1993

. . . Life goes on as usual here, except we had one heck of a late hot summer. We had 17 days in succession with the temperature between 42 [107.6] and 47 [116.6] degrees. And that was on my verandah too! Lord knows what it was in the sun.

Lost a lot of plants in the garden this summer. (They were some of the established plants. Fortunately I didn't plant anything this summer. Had a feeling it would be a sizzler.) Will put a few vegies and flowers in, in the autumn.

1 August 1994

. . . Life here is very quiet. Many people have left the town for various reasons, no opal, have opal but the price is lousy unless you have very good quality stone. I have, several times, threatened to sell up and leave as I am sure I could live a lot cheaper elsewhere. The price of water, of course is the main bugbear. I have let my garden run down to a mere skeleton of what I used to grow. I just concentrate on my chooks as they are the means of a few extra dollars with the selling of a few eggs . . .

8 December 1994

. . . Since your visit here—about two and a half to three years ago—we are still waiting for rain. It is now five or six years since our last good rains. I am now using 100 per cent town water as my rain-water has long run out.

The local area is looking very bleak and dry . . . as for my own backyard, am afraid you wouldn't recognise it . . . I am growing no more flowers or vegetables. I started four small patches of lucerne to supply greens to my chooks, but the result was a water bill of $80 for the month ($20 for the week) absolutely ridiculous for just one person in the house. Of course most of it went on the garden. So I reckoned it was time to re-assess the water usage problem as there was no sign of any rain, and decided to cut down on planting anything more and as the existing plants come to the end of their term put no more in.

Later on, I will probably put in a few plants indigenous to Australian conditions so they are more or less self supporting once they are established. Have got two young acacias and they are all right. The chooks are doing quite well.

Notes

[1] The soils in the Coober Pedy region are sodic, resulting in poor infiltration of water, impeded drainage and a high tendency to erode during intense rainfall. They are of low fertility, with free lime commonly present at depth, which on very alkaline sites may result in plants suffering from lime-induced nutrient deficiencies (Zwar, Beal & Oddermatt n.d.).

Some people feel the rain, but some just get wet

Betty Westwood

I am very attached to this country. I am concerned about the land. And you turn around and see the very old, dead and dying trees, and if you look into the future all you can visualise is that there will be nothing here at all.

For me there's something more to planting trees than just having concern. I've always had the feeling that I was part of the environment. Something drives me to do it. When I look at that big, old, ring-barked tree, dead and still standing, it's what gives me incentive. It gives me strength. If I can grow a tree, then plant it so a bird will nest in it, that's the ultimate happiness to me.

Something has to be done about the state of our land. We will never be able to bring it back, the way it used to be, but the least we can do is save and re-establish what we can.

Strathalbyn

Before European settlement, forests covered 10 per cent of the Australian continent, and woodlands covered 23 per cent. Some 50 per cent of those forests and 35 per cent of the woodlands have been cleared or severely modified (Bird *et al.* 1992). In 1840, John Robertson settled open woodlands at Wando Vale in western Victoria, and in 1854 he described the changing landscape resulting from opening the country to grazing sheep. His account epitomises the effects of agricultural practices that continue to have disastrous effects on the health of the land:

> *The few sheep at first made little impression on the face of the country . . . many of our herbaceous plants began to disappear from the pasture land; the silk grass began to show itself . . . the long deep-rooted grasses that held our strong clay hills together have died out . . . the clay hills are slipping in all directions . . . now that soil is getting trodden hard with stock, springs of salt water are bursting out in every hollow or watercourse . . . ruts, seven, eight, and ten feet deep, and as wide, are found for miles, where two years ago it was covered with tussocky grasses like a land marsh . . . the country will not carry stock that is in it at present . . . (Bride 1983).[1]*

Tree loss has been associated with almost every aspect of land degradation. Tree clearing for agriculture, and soil disturbance associated with grazing and cultivation, have produced massive soil erosion. Conservative estimates suggest that 43 million hectares or 13 per cent of our pastoral land is seriously degraded, with soil losses exceeding 20 tonnes per hectare annually. Retaining the existing trees in these landscapes and planting more trees are critical for re-establishing the fertility of degraded soils as well as for developing appropriate farm planning for sustainable systems of agriculture that help replenish the soil. Our production systems are not sustainable; we are mining the land with no care for the future.[2]

When All The Trees Have Gone

When all the trees have gone
No joyous song will greet the light
Or share its happiness all day
No bird wings home at night
When all the trees have gone

When all the trees have gone
No roots will hold the earth's thin crust
An age of weathered rock
Blown out to sea as dust
When all the trees have gone

No harvest time will come
No gentle grass that once forgave
Our greed; the desert soon
Will claim the land we failed to save
Now all the trees have gone

Betty Westwood

Photograph, Kay Johnson

'I only grow what is indigenous; I don't plant weeds.'

BETTY Westwood retired from nursing in 1981 and moved from Adelaide to Strathalbyn in South Australia, beginning her planting strategy in 1984. Betty is troubled by the ramifications of extensive land clearing, not only to the health of the land but also to the animal, bird and insect life that are part of the landscape. Her devotion to planting trees is partly inspired by the harmony of nature and an appreciation of relationships between the earth and living things.

Betty has a vision of what the landscape once looked like, and she uses existing signs and scars as a guide. She is a member of Trees for Life[3] and is passionate about the need to raise and plant indigenous vegetation. She collects local seed and propagates it in her backyard nursery, growing 4000 seedlings a year to be planted in the surrounding country. The joy of her work is not only in watching a sparse landscape grow and transform under a canopy, or a single tortured trunk become part of a stand of trees, but also in witnessing the encouraging effects that the trees have on the wildlife in the local area.

Betty

I am very attached to this country. I am concerned about the land. The 30 miles [about 50 kilometres] from this farm to the Murray River was once thick, thick bush. Now the bush is gone and the repercussions are damaging. I've seen the sand drift through this country in the winter time, the precious topsoil drift away. And you turn around and see the very old, dead and dying trees and if you look into the future all you can visualise is that there will be nothing here at all. There is a desperate need for planting and for protecting what we still have. I see this desperate need, but I think there is something other than this realisation as to why I am planting. For me, there's something more to planting trees than just having concern. I've always had a feeling that I was part of the environment. I was always very aware of plants and animals.

Nursing took up most of my working life, but I always hoped to be able to go back, as it were, to the soil. I am very aware of what I call the sanctity of the earth and when I did retire, I had the opportunity to work with it. I think that behind it all is an old Quaker exhortation: 'Thou hast a concern, therefore thou must do it.' Something drives one to do it and it gives me such—it's hard to explain—but it gives me incredible joy. Even when I am out here working, I am not conscious of my arthritis; I don't get as much pain, I'm not aware of it as much as when I'm not here. It does something to one. It's the healing—the soil is so healing.

I think that I am very fortunate to have this farm to plant on, although I would have found somewhere else to go.[4] I've always been a tree planter, but I've never had the opportunity to do it on this scale before. I am not the only one doing this. Lots of people are doing it. I propagate about 4000 plants a year, which are all given away except for those I plant. I really believe that you have to be very careful and very selective in what you chose to grow and plant. Everything that we have planted has come from our own local seed collection. They are all what would have once grown here, including she-oaks, wattles, eucalypts and little ground covers. I only grow what is indigenous. We must have the local seeds.

It's from working out here that I receive the inspiration to keep going. When I look at that big, old, ring-barked tree, dead, and still standing, it's what gives me incentive. Now I used to look at that tree and I would feel within me this terrible pity for it and

for what we did to this country. I think that just looking at that same tree many-a-time has kept me going.
It gives me strength.

I came here after retiring at the beginning of
1981. A couple of years later I started planting. The
first plantings were in 1984; in those days I could do it
on my own. Over several years, the rest were planted.
Then, one disastrous day, the fence broke down and
the sheep got in. How long they had been in there I
don't know, but they had eaten everything. I didn't
come near here for months—I couldn't bear it. Fortunately the trees were well rooted so they grew back.

All the recent plantings are a combined effort of
my friends and myself. My friends now do all the
heavy work, as I just can't dig and lift as I once did. I'll
show you what one friend made me because I can't
kneel any more. He made me this amazing instru-
ment. I take the tree and place it down at the end of
the tube, and then put it in the hole. I then put the
earth in and take the tube out, and there's the little
tree. This is my special planter.

We collect the seed, mostly from the existing
trees on the property, grow them in the nursery in my
garden, and then come out and plant them. At the
time of planting, we put a little water in the hole to
moisten the bottom, and then give them a drink after
they have been planted, and that's it. They don't need
anything else apart from newspaper placed around

*'They are beautiful, when you
look at those native lilacs, you
wonder why you plant any of the
other.'*

the base of the tree which suppresses the weeds for about a year. We have planted hundreds and hundreds of trees and ground covers. We don't count them, but there's more than you can really see, as there are lots of little plants—little low ones. We have a very high success rate. Very seldom do we lose a tree —very seldom. Occasionally you may lose a few from the odd rabbit or hare, but there really aren't many in these parts. Even more important than planting trees, is fencing off any remnant vegetation. Fence it off and let nature look after it. There just seems to be so few stands remaining of local vegetation. My godson, John Bradford, is fencing off the corners of the paddocks in order to establish planting. He's also fencing off the old trees for me, so that they can regenerate. There's a special, very big, old tree, just up here, and he's fencing it off for my birthday. We are making a bushland around the old trees. He said that if I had my way, the whole farm would be fenced off.

When we began planting, there were just a few old trees and a very sandy patch of ground. We started planting mainly to see if anything would grow. Of course, many of the trees are still very young, but in time it will sort itself out: some will die, and some will grow bigger, and in a few years you'll be pushing your way through the bush. To be able to restore vegetation, even to this degree, gives me the greatest joy.

However we cannot only think in terms of revegetating the land, or only planting trees. We have to plant habitat, not just canopy. We have to be aware of all the layers of vegetation, so we are not only looking at the welfare of the land, but also of the birds and animals in terms of where they may live. See the little Mallee tree that is starting to flower and the butterfly on it? Isn't it beautiful? That's really beautiful! It's so exciting to see birds and butterflies come back. That butterfly wouldn't be here if that plant weren't here.

If I can grow a tree, then plant it so a bird will nest in it, that's the ultimate happiness to me. This is why the old trees are so vital. They are habitat. I think that the older the tree, the more wonderful it is. It's the hollows—they are so important for the birds. You can't plant time. It's the same with the logs on the ground, too. Sometimes we drag them here. In fact, I tied two old boughs to the back of my car and dragged them up the road to spread up here as habitat. Isn't that a wonderful old log? You think of the age of that tree. I think that's the best log of all—it's wonderful! There are birds in it, too. I just have to look at that and it almost drives me to a frenzy!

Today, life seems to be all economics, but you just look at that skyline of old trees. It should so grieve people that soon they won't be there. There should be such regard for the health of our old trees.

'My generation don't think the way as I do. But I like to get out here.'

Photographs, Kay Johnson

Something has to be done about the state of our land. We will never be able to bring it all back, the way it used to be, but the least we can do is save and re-establish what we can. In doing this, the trees will eventually perpetuate themselves and we will be able to retain the unique sense of timelessness. We owe this to the soil and to the Aborigines who once lived here.

From a letter from Betty Westwood, 18 May 1998

Five years have now passed. The farm stands like an oasis in almost treeless country. There is no sand drift and sheep loss is very low.

'This is my special planter.'

My cousin, Diana Bradford, died and the farm was bought by a neighbour who generously supports my work.

Save the Bush, Department of the Environment, provided funds for fencing, and there is now a wide corridor enclosing remnant vegetation. With no grazing, native plants, including many grasses, are returning.

The Murray Darling Association has twice given me their Theo Charles-Jones Tree Award for revegetation work in Region 6.

Trees for Life continue to provide propagation material. The founder of Men of the Trees, Sir Richard St Barbe Baker, will forever be my inspiration. 'Save the Bush.' I recall the words of the late Professor Sir John Cleland pleading with Councils to save their roadside remnant vegetation. 'It costs nothing to preserve, and once destroyed, never, in all the ages, can it be restored.'

In addition to twice being a recipient of the Theo Charles-Jones Award, Betty Westwood has received a Civic Trust Award and the Order of Australia for her environmental work.

Notes

[1] This quote was contributed by Dr Rod Bird of the Pastoral and Veterinary Institute, Department of Agriculture, Hamilton, Victoria.

[2] Institute of Foresters of Australia 1989; Bird *et al.* 1992; R. Bird, personal communication 27 May 1993.

[3] Trees for Life is a tree planting organisation. It was established as a branch of Men of the Trees, which was founded in Kenya in 1922 by the English-born forester Dr Richard St Barbe Baker, who visited South Australia in 1981.

[4] The 'Bletchley Farm' of Betty's cousin, Diana Bradford, has become one of her main planting efforts.

Will we end up with a muddy palette?

Liz Fenton

On my property, 'Larapinta', I could do something about the future environment. I see tree planting, revegetation and caring for the land as a similar occupation to nursing— health and well-being of the whole community.

Over the years I've become more convinced that it is important to be planting locally collected material— indigenous plants. The same plant collected from different areas may look quite different and show differences in their ability to survive local conditions. I sometimes liken mixing indigenous plants to an artist's palette—when they are mixed together over large areas we loose the individual characteristics of the landscape.

I have paid attention to providing genetically diverse material with the plants that I've supplied over the last couple of years with a possibility for them to become seed sources for the future. Everything should be planted with a view of it being a seed source.

I've become aware of the vulnerability of the indigenous plants. I don't say that you shouldn't plant the others, but I think that we have a responsibility to ensure that through our actions the indigenous, or local, flora remains.

'Larapinta'

LIZ Fenton started growing trees for 'Larapinta', her property just outside Hamilton in the Western District of Victoria, and has since established a nursery to propagate local plants for the growing demand in indigenous farm plantations. Liz believes that for major property plantings it is no longer justified to use only exotic plants or native species that are not indigenous to the area. She is convinced that local plants have to be incorporated to ensure that bio-diversity within plant species is maintained. Liz's ambition is to establish seed sources for future planting schemes, and to retain the biodiversity within her local region.

Liz

It's the luck of the draw that we were born in Australia and not in any other country. If we were in Bangladesh, for example, what could we do? There would be little we could do as individuals to alter the landscape or preserve or restore the environment. It would be a case of survival. In Australia, we have these opportunities. As Australians we have a responsibility to the land.

I stopped nursing because I thought that we had gone a bit mad. We could preserve life but we weren't guaranteeing any quality further down the line, or a decent environment to live in. I decided that I was going to plant trees.

On my property, 'Larapinta', I could do something about the future environment. I wanted to learn about trees—growing trees and planting trees. I used any opportunity to work with people who were working with trees, and it's gone on from there. I see tree planting, revegetating and caring for the land as a similar occupation to nursing—health and well-being of the whole community. If I looked at it purely as a job, it's one which I find very satisfying, and in many ways I'm totally in control of what's going on. Often with nursing you're not in control of how well you do your job.

I started growing trees for my farm. Over the years I've become more convinced that it is important to be planting locally collected material—indigenous plants. Experience is showing that seed collected from local plants of a particular species

generally performs well in the short and long term. The same plant collected from different areas may look quite different and show differences in their ability to survive local conditions. To talk about indigenous plants for farm plantations has almost been in front of the demand for native plant stock, but the demand is following. I've contract-grown for local farms, and I'm finding there's an increasing number of farmers who are paying attention to using local plants for shelter belts. Up until recently no one was really interested in that. Two years ago I sold about 5000 plants, this year it's been closer to 30 000 and the demand is increasing.

Sometimes when I'm talking to farmers I compare the Redgums [*Eucalyptus camaldulensis*] with Merino sheep. They both grow or survive across most of Australia, but no sheep breeder would consider taking a superfine Merino from the Victoria Valley in southwest Victoria and put it up in outback Queensland and expect it to thrive. Maybe we shouldn't expect Redgums or other plants from one area to thrive in a vastly different situation! I sometimes liken mixing indigenous plants to an artist's palette - when they are mixed together over large areas we loose the individual characteristics of the landscape. If you mix all the colours together you loose track of each one and its particular vibrancy and end up with a muddy brown or grey.

Five-year-old trees in the Quarter Centennial Planting.

Originally, I wasn't truly indigenous with the range of plants I grew or planted, but I think we've got to be looking at local species if we are going to maintain our genetic diversity. Last year's planting has been strictly indigenous and I've kept records of the seed sources. As a matter of course I now provide detailed information to clients when they purchase either trees or seeds about the origin of the seed and the number of parent plants it was collected from. It's pleasing that some people are coming back and commenting about different characteristics within the same plant species. The Redgums that we planted along the streams on my property have been grown from seed collected from more that fifty different remnant parent trees. Ultimately I see that they may become quite an important seed production area, offering genetic diversity for this region. This diversity increases the chances of the species surviving. The seed from each of those old parent trees is being individually grown and is identifiable in the nursery. I provide that diversity of seedlings to my clients. Therefore, if somebody wants a box of Redgums, there's about fifty in a box and no two trees are from the same parent.

I have payed attention to providing genetically diverse material with the plants that I've supplied over the last couple of years with a possibility for them to become seed sources for the future. Now, I don't know whether it is or isn't important to worry about such detail, but if it is important we've got to act right now. If it isn't important well there's no harm done. But every year that we leave it, the greater the risk of losing the diversity. There are local scientists and farmers who estimate that there is between a 2 and 10 per cent loss of our big trees on an annual basis, just through wear and tear, and for ground cover it is probably far greater. We've lost huge percentages of the ground cover and understorey plants and in many cases particular species have become rare or extinct. Everything should be planted with a view to it being a seed source.

At the moment I know that a number of people are collecting seed from plantations and growing the next lot from the plantation-collected seed. I suspect a lot of the seed has originally come from one or two parent plants, so you immediately run the risk of inbreeding and consequently, in some situations, reduced viability and vigour. If we get it right now, and start planting with local plants and keep records, these plantations will become increasingly valuable as seed production areas. They could be the backbone of a growth of indigenous plants later on. It's important to preserve what we've got—whatever we re-create is only going to be partly as good as what was originally there.

The people who are most supportive of my work are often the older farmers. They're the ones who have generally retired and reflect on what's going on, and they look back at what the land was like when they were young and just starting out. They know they put a hell of a lot of effort into clearing, and any criticism they make is about themselves. Whereas the next generation are sometimes a bit protective of their ancestors and are afraid to see them criticised. It's the older people who have seen the land change, and it's happened over such a short time.

I grew two or three thousand trees for the first year, and a number of people from the city approached me to see if they could be a part of the tree

'I get the greatest thrill collecting the guards from the trees that don't need them anymore.'

planting. It was the year before the bicentennial and so we said, 'Well, let's do something called a quarter centennial planting, because what we will be planting now will be reaching maturity in a couple of hundred years'. The people have come back and their little kids have seen the plantings grow much bigger than themselves. So these little kids who were 12- and 18-month-old babies sitting in the mud, playing with trees, have come back and found koalas in the trees which are less than four years old.

These young trees have been of help to farm animals, or animals with special needs in critical weather conditions. We've had premature calves, premature triplet calves, and I would say they survived thanks to going into treed shelter areas. The plantations are a life saver and help reduce stress for newly shorn sheep.

The trees have increased the numbers of birds and native animals. You used to see them intermittently, but now a lot of them hang around for a few weeks during each visit. For two years running I had

Black Tail Wallabies for five or six weeks over January and occasional kangaroos, and we are seeing echidnas again. It is also the hollow standing trees that are important. Lots of parrots are using them at the moment, sometimes bees and ducks use them.

I don't remove logs that are hollow or rotting down, because I think they are valuable as habitat. There are bats, insects and the whole range of little things on the ground that need the rotting material to live in. They also provide habitat for reptiles. The tiger snakes are a bit scary. I sometimes wish they were a bit noisier—it would be reassuring if they roared. I've become aware of the vulnerability of the indigenous plants. I don't say that you shouldn't plant the others, but I think that we have a responsibility to ensure that through our actions the indigenous, or local, flora remains. If we pay a little bit of attention to the seed source at the beginning of a planting project, we ensure that the area becomes a valuable seed-collecting asset further down the line. It's only a very small aspect of a whole project, and perhaps the tedious one,

but it's the one that stops you from getting started. At this stage I think it is sometimes a little bit too hard for people to take on board, but it is something really important. To know that your actions can have a positive impact is exciting!

Lanark, a sustainable farm for the future

John and Cicely Fenton

*When I went to 'Lanark' it was absolutely bare and all the
water had been drained. So there were no trees and no
water, but it was good grazing country. I'd been given a
message by the family that it was our property and I
needed to look after it, so around the same time I started
to plant a few cypress trees, but they all died.*

*I suppose I became involved in this work because I saw the
property a little bit differently than a lot of farmers view
their land. It became an all-embracing hobby—planting
trees. I'm really happiest planting trees. And when they
have grown I put my ear to the trunk and I can hear the
wind whistling in the top of it, and I can hear the
movement, and there is some sense of the tree being
alive—that I'm alive.*

*The 1967–68 drought was one of our big downs, when the
water actually dried up. But the drought spurred me to
build a bigger dam at the front of the house. We put back a
fair bit of water—virtually all the water—and soon the birds
returned. One half of the increase in bird species was due
to the wetlands improvement and the other half to the
shrub and tree planting.*

The Western District of Victoria, also known as the Green Triangle, contains Australia's prime wool growing country. Before European settlement, the landscape was seen by the local Aboriginal people as broad open valleys, shallow lakes and swamps, and covered by open woodlands and savannah vegetation.[1] The significant agricultural value of the land was recognised by Major Thomas Mitchell when he and his party passed through the area in September 1836. His glowing reports of the area led to its rapid occupation by pastoralists:

> *This land is, in short, open and available in its present state for all purposes of civilised man. We traversed it in two directions with heavy carts, meeting no other obstruction than the softness of the rich soil; and in returning over flowery plains and green hills fanned by the breeze of early spring, I named this region Australia Felix, the better to distinguish it from the parched deserts of the interior country . . . flocks might be put out upon its hills or the plough at once set to work in the plains (Mitchell 1838).*

There was little clearing during the early pastoral phase, but the gold-mining boom of the mid-1800s and the development of the pastoral industry after alluvial gold deposits were exhausted resulted in large-scale tree clearing. The dramatically increased population needed firewood, pit props, housing and railway sleepers. After the mining boom, the Land Act of 1884 required that new selectors ring-bark or fell trees in order to obtain freehold title to their blocks. By 1903, the district south of Hamilton had been so cleared that there was an emerging need for establishing shelter plantations (Bird n.d.). Remaining trees were, and continue to be, affected by fire, insects, disease, soil compaction by grazing stock, cultivation, rising water tables and old age. Today, open windswept landscapes characterise the Western District, scarred by soil erosion and increased water salinity.[2]

'Lanark' is a 730 hectare wool growing property on the edge of the Crawford River catchment near Branxholme, approximately 25 kilometres southwest of Hamilton. Its landscape consists of broadly undu-

Photograph, Lindsay Stepanow

lating basalt plains, with extensive naturally occurring wetlands. The soils are grey-brown silty loam over buckshot gravelly loam and clay. The average annual rainfall is approximately 700 millimetres. Historically the property would have supported open woodland including eucalypts, wattles and tea-tree.[3] Before John Fenton inherited 'Lanark' from his father in 1956 it was an out-paddock of a much larger property. It had been totally cleared of all remnant vegetation and drained of its natural water bodies to maximise the production of wool and fat lambs. 'Lanark' was open to the elements, isolated from wildlife and almost entirely devoid of natural habitats.

John and Cicely have worked to progressively reinstate the wetlands and to plant approximately 80 000 trees including indigenous plantings for shade, shelterbelts and habitat, and commercial agroforestry and woodlot plantations. John describes agroforestry, or farm forestry as 'not pure forestry, but rather an endeavour by a land owner/manager/ farmer to grow selected species of trees on his land with the intention that most will be managed to produce high quality timber for sale or for on farm use which, on 'Lanark' will be over a rotation of 25 years to 125 years' (Ruddock 1995). He and Cicely have worked with scientists, field naturalists and education institutions, including Dr Rod Bird at Hamilton's Pastoral and Veterinary Institute. Dr Bird and his colleagues contend that the long-term development of up to 15 per cent of the entire property in shelter/timberbelts and woodlots will reduce salinity and erosion and produce a more sustainable agricultural system without affecting farm incomes. They argue that if 10 per cent of an agriculturally productive area is devoted to trees, the shelter provided will increase productivity by approximately 15 per cent.[4]

Trees and water have not only provided practical and aesthetic benefits to the property but have also allowed the development of a diversified farm income strategy. Extending production from grazing into areas like agroforestry has increased the property's profitability. Plantations include commercial softwood and hardwood plantings managed for the production of high quality clearwood. Pine (*Pinus radiata*) and eucalypt clearwood logs are expected to be harvested within 20 and 35 years respectively for domestic and export markets. By 1994 nearly 19 per cent of the property was not being used for animal, crop or pasture production but was dedicated to farm forestry, habitat plantings and wetland, amenity plantings and shelter plantations. Jeff Tombleson of the New Zealand Forest Research Institute anticipates that by 2009 the 6 per cent of the property dedicated to farm forestry could be generating up to 70 per cent of the net farm income. These efforts were recognised in 1994 when John and Cicely received the Stihl Australian Forest Grower Victorian Tree Farmer Award followed by the Australian Forest Growers National Tree Farmer Award.

John and Cicely are pleasantly surprised to be enjoying the results, which they thought would be the pleasure solely of the next generation. Examples of truly integrated broad-acre commercial farming with conservation are few, but it can be achieved in one working life. In November 1994 'Lanark' was selected for the first wild release, on private land, of the Eastern Barred Bandicoot, which was once widespread throughout the state but is now Victoria's most endangered mammal. The Department of Conservation and Natural Resources released the colony of bandicoots in a bushland area that John and Cicely planted soon after they were married.

Photograph, David Fenton

JOHN and Cicely Fenton are pioneers in developing a balance between ecology and economy on their wool growing property 'Lanark'. They have integrated grazing and agroforestry (farm forestry) with wetlands, shelters and habitats for wildlife, to create a dynamic ecological environment. Reinstating large areas of water and planting trees has in the past been contrary to accepted local farming practices. However, they are now valued as beneficial to farm productivity as well as to environmental health. Tree shelter reduces the stress to stock from cold wind and rain, and improves crops and pastures. Establishing habitat for birds, which are natural insect predators, helps to eliminate the need for pesticides. The natural filtration system of healthy wetland areas provides cleaner waters for stock and homestead irrigation.

Assessments are continually being made by the Pastoral Research Institute to determine the effects of the Fentons' management practices on the ecology of the property. An increase in bird species is one indicator that environmental health has dramatically increased. Murray Gunn, a local Hamilton ornithologist, has kept consistent records of bird sightings on 'Lanark' for over forty years. They show a direct relationship between the reinstatement of wetland habitats, tree and shrub plantations and an increase in the numbers of bird species belonging to those habitats.

John and Cicely's efforts have not only created a legacy for their children but have also established the importance of native trees and wildlife on a productive property. This is contributing to the development of a new farming attitude for southeast Australia.

John

Looking back, I often wonder why the hell I didn't sell the property and buy a watered and treed farm, but I suppose I was charged with some family responsibility, although 'Lanark' had only been in my family since 1938.

The 'Lanark' property had been owned by my father. It was an unimproved portion of 'Bassett

Station', in the southwest corner of Victoria. Bassett was an old property that dated from the time of the first white settlement in the area and, in contrast to 'Lanark', retained the pleasant old homestead down in the valley—a select site on the most fertile piece of soil with permanent water and a lovely garden. I'd grown up in similar conditions living in Hamilton. My father died when I was nine. I lived with my mother in a very comfortable house, and I'd worked on the land elsewhere, but on very fortunate properties—on wealthy properties. And when I went to 'Lanark', it was absolutely bare and all the water had been drained. So there were no trees and no water, but it was very good grazing country.

The two bedroom house on the farm had been built at a cost of £1400 straight after World War II, so here I was in this bloody horrendous hot-house. I'd been given a message by the family that it was our property and I needed to look after it, so around the same time I started to plant a few cypress trees, but they all died. I then became engaged to Cicely and my prospective father-in-law came down for a visit. He lived on 'Vasey Farm', where Cicely had grown up and which my son Paddy now manages and part owns in the Red Gum country [*Eucalyptus camaldulensis* predominates]—a treed property which is very pleasant. Cicely's father was of English stock, and a person with an intuitive feeling for pleasant places, and he turned around to me and in his cultured voice he said, 'If you want my daughter to stay with you on this god-forsaken place, you'll have to do something about it. This is a bastard of a place!'

I suppose I became involved in this work because I saw the property a little bit differently than a lot of farmers view their land. Most farmers are brought up on the land and they never leave the property. It would be the same as if people in the city were all born in the same house and stayed there for your lifetime, and the house passed from generation to generation. Your view of the world would be that house. I had also befriended a band of people other than farmers, who certainly saw the world around them differently, and I became more and more interested in some of their ideas and in the bird life and landscape, so I sought those people out. That in turn makes my family think I'm sometimes a bit of a pain

Photographs courtesy John Fenton

'Lanark' homestead 1963

**John and eldest son Johnny dig
test holes for the dam, 1967**

Photograph, Lindsay Stepanow

'Lanark' 1995

in the neck, and it certainly upsets the bankers and the farm management consultants. They say, 'Garbage! What's that going to do for your cheque book?'

It doesn't worry me that I'm a bit out of kilter with everybody. People would say to me, 'You bloody idiot, you're blocking up another swamp', or, 'Why don't you drain the dam thing?'

The advisory man and the manager of the old Ballarat Trustee Company that ran my father's estate, said to me one day in 1959, 'We want you to drain that 60 acre [about 24 hectares] swamp'.

'I have just increased the size of it.' I said.

'You what? I didn't see any account for that.'

'Well it didn't cost much, I did it myself.'

'I want you to drain it.'

'What for?' I asked.

Agroforestry plots

Photograph, Lindsay Stepanow

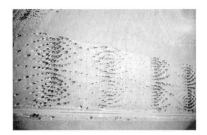

Photographs, Lindsay Stepanow

'Because we can get a government subsidy to do the work.'

'Well, that's no bloody reason to drain it!' I told him.

When the cypress trees had died, I was fortunate to hear an old chap from the forest commission nursery at Wail in the Wimmera talking on the wireless, so I went up to see him. At about the same time I visited the NRCL [National Resources Conservation League] nursery down at Springvale, and I got a very good reception from both. Fortunately those people knew what they were doing in those days, and they gave me the right recipe and generally the right trees, and in amongst them were some wattles which grew particularly quickly, and that fired me up. It became an all-embracing hobby, planting trees. But I think if they'd have given me another batch of pot-bound cypresses and they'd all died, I'd probably have given it away. These were the things that initially spurred me to improve the farm.

There are about thirty-five pine trees along the railway line that were planted during the Depression, about three hundred pine trees that my father planted in 1939, and there are two very small Sugar Gum [*Eucalyptus cladocalyx*] plantations on the property. Every other tree that I have the opportunity or privilege to walk under I have planted myself or an extremely close person has planted. So there's a sort of a personal friendship with every tree and you know that every tree that you've planted is going to produce some good.

I'm really happiest planting trees. Funnily enough not raising them, not putting seeds in the ground and growing them and potting them, but just planting trees, six inch trees. I get them from all sorts of people who know more about propagation than I do. I love to buy a box full of trees and go out and plant them. And when they have grown I put my ear to the trunk and I can hear the wind whistling in the top of it, and I can hear the movement, and there is some sense of the tree being alive—that I'm alive. My 2-year-old grandson is a fast learner, and he loves to play the 'ear to the tree' game.

Cicely has the problem of doing the property books and fighting budgetary constraints so we take a slightly different approach. Cicely says, 'You won't plant any more trees'. And I say, 'Well, if I can't plant more trees, there's no point in me being here'.

Cicely

I'm not ready for more trees unless there's a big improvement in the quality of pasture management and pasture renovation. We already have nearly 10 per cent of our property in trees. I don't see that you can go on putting more and more trees on the farm and expect to carry the number of sheep that you need to be commercially viable. And you can't do that unless you have good pastures underneath. It's as simple as that.

In the long term having 15 per cent of our property covered with trees is a goal; when you start getting returns from the agroforestry maybe we could increase it to 20 per cent. I suppose you could do it now but the money is not there, so it's one of those things that has to be part of a long-term plan. Unfortunately, when you get a recession everybody stops doing the things that they consider aren't necessary, and trees are only an extra thing in a farmer's mind, they don't make them any money—at least not immediately.

I'd be much happier about saying yes to planting more trees if we didn't have the debts that we are carrying. If you're going to retain the farm, or any farm, in this day and age, you can't afford to be carrying too much debt. Maybe I'm the only realistic one in this household, the others are too airy-fairy. It's a good dream, having more trees. It's a very necessary dream—but it has to be realistic too.

John

I attended a major tree conference in Melbourne in 1980, and I suddenly realised there was a great heap of people, both official and unofficial, from all over Australia who were doing exactly the same as me, and in fact wanted to hear about what I'd done. So after this meeting we formed a branch of the Men of the Trees, and I thought that was a important thing to do.

Photograph courtesy John Fenton

Last of the water on 'Lanark'
during the 1967/68 drought

I became involved with them, in fact initially I was the only broadacre farmer, and we used to meet every couple of months.

Shortly after that the 90-year-old English-born forester Dr Richard St Barbe Baker, founder of the Men of the Trees, came to Melbourne. I remember listening to an emotional interview with him on ABC radio during the four hour drive to Melbourne to meet him, I still remember my sense of expectation and excitement. A public meeting was held the next day in Wilson Hall at Melbourne University, which was chock-a-block. We had to wheel the old man in, and he rose out of the wheelchair halfway up the hall, and said, 'Stand back, I'll walk'. It was like dealing with God. [5]

At the dinner the night before, he'd insisted that everybody sit next to him in turn. He put his hand on my knee, and said, 'Young man (me being nearly fifty) do you like trees?' And I said, 'Yes sir, I do'. And he grabbed my bloody knee in a vice-like grip with his bony old hand and he said, 'But do you plant trees?' And I said, 'Yes'. 'How many?' And I told him, 'Thirty thousand'. And he said, 'Good, good, good', and went to sleep. So he thought my efforts were satisfactory enough for him to have a bit of a snooze. That gave me a bit of a boost because I had all his books, and people were saying I was doing too much, but he seemed to be saying I could do even more.

There's a word 'procreation' and I suppose the biggest joys over the past thirty-five years have been in the breeding of children, animals and trees. There's more fun in breeding children than there is in breeding sheep, however satisfying that can be, but it's a great satisfaction in a farmer's life to breed things. Seeing the birds breeding is one of the satisfactions of putting water back on the farm. Putting water back is a one-off cost. The outlay is relatively small, and thereafter you get this continual filling of those areas.

The 1967/68 drought was one of our big downs, when the water actually dried up. The drought was a costly exercise in a variety of ways and it was also very, very hard work. I feel it sort of started a deep degeneration of the human resource of the countryside in the Western District. It stripped the countryside of labour—I think it was necessary labour, and not only labour, but company, and that's never

'I'm really happiest planting trees.'

come back. So still today, farmers are living fairly lonely and very hard-working lives under a lot of different pressures. But the drought spurred me on to build a bigger dam at the front of the house— Lake Cicely.

The farm used to be very hot, without shade and water. I mean even in the Western District, it's bloody hot in the summer, and we very rarely have a holiday by the seaside because of the bushfire danger. There was a remnant drained swamp on the place and it took very little expense to reinstate it. So the wetlands were useful as a fire retardant, as well as being lovely to look at, attracting birds and providing somewhere to swim. I suppose I already had some interest in bird life. I'd been influenced by a family friend, Murray Gunn, who has a great appreciation of natural history and of birds. But we went overboard in a way, putting back 145 acres [nearly 60 hectares] of wetlands on the then 1715 acres [nearly 700 hectares] of farmland. We put back a fair bit of water —virtually all the water—and soon the birds returned! Murray Gunn's figures show that there were 43 birds species on the farm in 1952 and today there's 156.

Half the increase in bird species was due to the wetlands improvement and the other half to the shrub and tree planting. World Wide Fund for Nature researchers tell us that the diversity of species and the number of individuals of each species on an area of farmland is directly related to the health of that given piece of land. Clearly it's an important guide.

'Lanark' is presently 2000 acres [about 800 hectares]. The 7 per cent of wetlands are only filled to

capacity on average once every eight years, depending on the seasonal rainfall. We have approximately 9.5 per cent of the property in trees and shrubs, 6 per cent in agroforestry, 2 per cent in shelter habitat plantations and 1.5 per cent in the garden around the house and arboretum. In my lifetime we hope to achieve 20 per cent of the property under tree cover: 10 per cent planting of indigenous trees, shrubs and ground flora for shelter, shade and habitat—for humans as well as birds and animals—and 10 per cent planted with commercial species—agroforestry and commercial woodlots. The 10 per cent of indigenous planting can also include limited harvesting for specialty timber and wood for farm use, as well as providing a seed-collection resource for revegetation projects. That's what I'm aiming for.

In order to accomplish these aims and build on what we have done over the past thirty-eight years we are developing three types of areas each with its own special conditions on 'Lanark'. One area is primarily for vegetation. Parts of these areas are fenced out forever for ecological benefits and others are fenced but stock have periodic access for management of the indigenous ground flora which had been grazed by native marsupials. You can graze those stock access areas from February until it gets wet without doing untold damage. Then there's the agroforestry area where the stock are grazing under the maturing trees all the time, and the trees themselves are harvestable as a source of income. Then you do the same with water. You have water bodies that are fenced out and the margin is permanently protected; some are

'There is a great joy in seeing the water and the dams fill.'

protected seasonally, and other areas that we call water meadows have stock grazing through them.

And then on the farming side, we need to differentiate between land that can cope with being cropped and land that can not. You have those areas that are partly native pastures that are looked after accordingly. They are not fertilised and are located in the catchments of the wetlands. You have other areas which are well maintained improved pastures for grazing and possibly areas set aside for some exotic crop. In the future we may well be growing crops that we don't know of yet.

The major water and indigenous plantings form a terminus point at 'Lanark'. It is one of the property's most important features. It's the heart of the design. This terminus leads out into a number of shelter belt and agroforestry plantings. The plantings are all interconnected for stock and human shade and shelter. The water area affords fire protection and visual beauty for humans and serves wildlife as well, especially the birds. The terminus is very sheltered and as you extend out into the property you reach

Agroforestry is the integration of forestry and agriculture on the same land.

less sheltered environments. This and other water-plant areas on 'Lanark' become contact points for wildlife and, when connected to roadside plantings, will have a broader impact on the ecology of the region. If every farm acted as one of these terminus points connected to the roadside indigenous vegetation, farmers would be spinning a living web, a living network across the country. Imagine this being a result of a farmer's part-time work.

From letters from John and Cicely Fenton

12 May 1998, from Cicely

In my opinion, conservation farming is becoming more difficult to be a part of as commodity prices and seasonal conditions are beyond our control. David has the ability to do things right and well if there were not the financial restraints on him that there are. I am also of the belief that farmers need five good years to be able to lift some of the financial

This terminus leads out into a sequence of shelterbelts and agroforestry plantings.

Photograph, Lindsay Stepanow

Cicely tags one of the Eastern Barred Bandicoots released on 'Lanark' in 1994 by the Department of Conservation and Natural Resources.

burdens that are on their farms. Not one or two good years, and believe me five in a row is unheard of!

This year is away for a good start weather wise, but who knows what is around the corner.

14 May 1998, from John

In one sense I still have an optimistic outlook. My family has embraced the values and ideas that I have striven for. Great joy comes with the opportunity to share my thoughts, particularly with young people. On the other hand I have a deep and sad feeling of pessimism and even cynicism regarding the state of the natural environment in broadacre Australia. Despite Landcare initiatives,[6] we are still destroying 600 000 hectares of bush per annum. There is no system in place to put even a basic value on the nations Natural Capital, and the financial institutions either don't understand or don't want to understand this, or address it.

At this time in our evolution as a nation, Australia generally has no true understanding of the fragility of the soil and water base of our farmland, and more than one politician and bureaucrat have dismissed my pleas as emotional nonsense. Rather than call for farmers to lift production to counter the cost/price squeeze, we should maximise the production for each unit of output, where possible value-adding to the product on the farm or in the region. The small number of 220 000 rural managers and farmers must receive more for their products, so as to allow a lessening of damage from an unsustainable activity on the land.

In south-west Australia the Australian Bureau of Agriculture and Resource Economics reports that the average farmer property returned a profit in 1996/ 1997 of only $131 - this was not per acre, but the net profit for the entire average property. I am convinced that a way must be found to return more to the farmers for their products. The nation must support farmers. If not, the structure of the rural community will fall to pieces. Small villages and towns are already virtually closed down, and if this happens widely throughout the country who will be left to put out the bushfire?

I now know that our vegetation program at 'Lanark' will not be completed by me, but be an ongoing activity of our son David and his son Will. The aim is for 40 per cent tree, shrub and ground cover, and as a New Zealand Agroforestry researcher says, if one half of the 40 per cent is planted with well-managed commercial timber species we will produce income that will sustain the expensive and destructive hobby of farming cloven footed animals at 'Lanark'!

We are now running farm tours to augment the property's income. This allows people who are concerned about environmental and sustainability issues surrounding the nation's agricultural systems to experience what we are trying to achieve. The more that people gain an understanding about new directions for agricultural systems that value the nation's Natural Capital, then the greater the voice will be to help create support for the necessary change to occur. Let us share with you our knowledge based on living and working at 'Lanark' for over forty years.[7]

John shares his knowledge with a group of students from RMIT University

Notes

[1] The savannah or open woodlands of the western plains
supported Kangaroo grass (*Themeda australis*), Wallaby
grass (*Danthonia spp.*), Weeping grass (*Microlaena
stipoides*), Blown grass (*Agrostis avenacea*), and Tussock
grasses (*Poa spp.*). As a result there was little need for
clearing during the early pastoral phase (Bird n.d.).

[2] Half of Victoria's crop and pasture lands are affected by or
at risk of land degradation (tree loss and associated soil
salinity, water and wind erosion, soil acidification, soil
structural decline and nutrient degradation) (Bird *et al.*
1992).

[3] The woodland vegetation community would have included
Acacia melanoxylon (Blackwood), *Acacia mearnsii* (Black
Wattle), *Allocasuarina verticillata* (Drooping She-oak), *Bur-
saria spinosa* (Sweet Bursaria), *Banksia marginata* (Silver
Banksia), *Eucalyptus ovata* (Swamp Gum), *Eucalyptus
viminalis* (Manna Gum), *Leptospermum lanigerum*
(Woolly Tea-tree) and *Leptospermum juniperinum* (Prickly
Tea-tree). The understorey species would have included
Themeda triandra (Kangaroo grass), *Stipa sp.* (Spear
grass), *Danthonia sp.* (Wallaby grass), *Poa sp.* (Tussock
grass), *Helichrysum sp., Kennedia sp.*, and a diverse range
of other small plants and groundcovers (Hay 1993).

[4] Such planting could achieve a 50 per cent wind speed
reduction, reducing stress on livestock and pastures.
Excessive heat and cold significantly increase the energy
requirements and expenditure of stock, diverting energy
away from productive functions. Indeed the combination of
strong cold wind and rain can double an animal's energy
maintenance requirements. Adequate shelter can reduce
the mortality of lambs by 50 per cent. By reducing ground
wind speed, shelter can reduce moisture loss through
evapotranspiration and increase relative humidity,
resulting in greater pasture production and quality
(Bird *et al.* 1992, Sinatra & Jones 1988).

[5] Dr St Barbe Baker died in 1982 in Canada aged ninety-two.

[6] Landcare is a multi-disciplinary, Australia-wide community
incentive program to promote and achieve more sustainable land and water management. Landcare tries to
balance economics and ecology, productivity and
resource protection, and contributes strongly to community development. Organisations and programs that helped
shape Landcare from the early 1980s include the Soil
Conservation Authority's group conservation projects, the
Garden State Committee's farm trees groups and the
Salinity Bureau in Victoria; Western Australia's system of
Land Conservation District Committees (LCDCs); the New

South Wales government's drive for Total Catchment Management; Tasmania's Private Forestry Division; the South Australian Soil Conservation Boards; the nationwide organisation Greening Australia; the National Soil Conservation Program; and the Australian Trust for Conservation Volunteers. In particular, the LCDCs in Western Australia with their statutory foundations, and the Victorian farm trees groups, created opportunities for local communities to work independently to address environmental problems. The term Landcare originated in Victoria through an initiative of Joan Kirner, then the Minister for Conservation, Forests and Lands, and Heather Mitchell, president of the Victorian Farmers Federation, who conceived the formation of a highly autonomous network of groups involved in land restoration across Victoria. The first group formed at Winjallok near St Arnaud on 25 November 1986. After twelve years, eight as a nationwide network, the Landcare movement consists of over 4000 community groups, mainly rural and composed of farmers and other land owners. People in Landcare are increasingly focusing on whole catchments and regional themes, rather than confining themselves to their own properties. Many landcare groups have amalgamated into loosely bound but highly task-oriented regional networks. By involving as many landowners as possible, creating an environment of trust, emphasising local planning and decision making, developing a fostering rather than a leadership role for government and promoting ready information-sharing between scientists and land managers, a highly productive era of land restoration is well under way. The wide acceptance of Landcare has led to close partnerships with industry, especially through a commercial and fundraising arm of the movement, set up by the federal government and known as Landcare Australia Limited (Marriot *et al.* 1998).

[7] Information about tours of 'Lanark' is available through the authors, or at 'Lanark' on telephone 03 55786243.

The next generation

David and Paddy Fenton and Amanda Fairbairn-Calvert

The trees have been a major part of Dad's life, so therefore it's had to influence us kids - and it has influenced us - I mean you can see that.
I don't think there will ever be salinity problems on 'Lanark' because Dad faced up to the problem so many years ago. It's all about caring for the land, isn't it? That's why the trees are so important.

David Fenton, 'Lanark'

To me it would be just perfect to have a farm where every paddock is enclosed with trees and their shelter and ecology are working together.
Our knowledge of caring for the land has been part of our upbringing - a part of everyday life. But Mum is the one I have to thank for giving us the opportunity to take on 'Vasey Farm', as it belonged to her parents - my grandparents. Yep, she has been a great help in getting us started - and Grandad's love of the tall timber was a bonus, you can see that in the landscape.

Paddy Fenton, 'Vasey Farm'

Being brought up in the environment at 'Lanark', which was so sheltered, here I really noticed the wind, the cold and the fact that you couldn't just look out and see trees and walk amongst them.

Charlie thinks a target of 10 per cent tree coverage for the farm is not beyond the realms of the imagination. On our property that would be 500 acres of trees.

Amanda Fairbairn-Calvert, 'Banongill East'

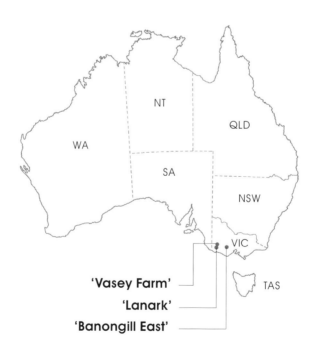

'Vasey Farm'

'Lanark'

'Banongill East'

THE passion and commitment that both John and Cicely Fenton have given to improving their property 'Lanark' have developed an agriculture philosophy in the family. It is expressed through the efforts of the Fenton children, all of whom have until recently remained in farming. Although Johnny, the youngest brother, sold his South Australian farm in 1992 due to economic pressures, David remains on 'Lanark' as the property manager; Patrick (Paddy) moved to 'Vasey Farm' northwest of Hamilton, where Cicely grew up; and Amanda married Charlie Fairbairn-Calvert and moved to 'Banongill East' on the eastern edge of the Western District.

Growing up on 'Lanark' has influenced the Fenton children in their attitudes towards establishing more appropriate pastoral systems with landcare being a primary component in their farm management plans. The Fenton legacy is a tribute to the perseverance of all farmers who are working to improve agricultural practices whilst having to battle against often uncontrollable events including financial uncertainties, fire threats, rabbit infestations, drought, cold and wet, changes in government policy, and land degradation. The Fenton legacy is also an important demonstration in developing environmentally and economically sustainable agriculture—a vision to keep landholders on their land whilst maintaining a greater respect for the land.

DAVID FENTON

David Fenton manages 'Lanark' and its 8000 sheep, as his father's main concern is the maintenance of the trees. David aims to raise the productive standard of the farm through improved breeding program for the stock and upgrading the pastures on the property, while maintaining and extending the agroforestry. He sees that 'Lanark' now has the challenge of becoming a sustainable property.

David

The trees have been a major part of dad's life, so therefore it's had to influence us kids—and it has influenced us—I mean you can see it. For instance, the knowledge of the wildlife on the farm has certainly rubbed off on me. It's quite funny, we might be sitting in the beach house at Robe in South Australia with friends of mine over from the southeast, and a yellow winged honeyeater or a little wattlebird will land in the tree out the front, and I say, 'Oh, look, it's a little wattlebird' or a 'firetailed finch'. My friends just burst out in hysterics. They wonder if I'm bullshitting or whether I actually know it. So now you find, if we're driving along the roadway somewhere, or we've stopped, or we're having a barbecue in the middle of the scrub, and a bird comes, they'll say, 'Oh, David will know'.

Dad would plant the whole place out in trees if he could, but in order to exist you've got to remember which side your bread is buttered on. The livestock, that's solely my responsibility. Over 80 per cent of 'Lanark' is under pasture, and trees help pasture improvement. I mean eventually when the tree gets really big you lose a bit of grass production immediately underneath, but further out in the paddock your grasses are growing better because of reduced wind stress, so it evens out. I mean, they've proven at the Pasture Research Institute in Hamilton that the loss of production through 10 per cent of the property being allocated to correctly sited and selected species of shrubs and trees is zero.

The trees help. When winter is upon us, the trees shelter the pastures, your ground stays warmer a little bit longer, and you get a bit of extra growth in your grasses. If I won a million dollars in Tattslotto, it would be great, I'd just plant the place out straight away. I believe that 15 per cent tree planting is what we've got to aim for and it should be able to be maintained without cutting back in livestock numbers as long as we do some work on the pastures.

Our pastures on 'Lanark' need to be improved and I have already started improving them to compensate for the extra mass of trees. I'll have to use a little bit of fertiliser under careful guidance. It won't be put out the usual way, that is every year. It will be put

David prunes trees in one of Lanark's agroforestry plots.

out once a soil test is done showing when it is needed, and keeping a close eye on the effects on native
pastures and wetland areas. By putting out a small
amount of superphosphate you can get added growth
on your Clovers and your perennial Rye Grass and
things that are already present. I mean, the Department of Agriculture has looked at my pastures, and
they are adequate to grow adequate wool, but for
good wool production they're not the best pastures.

I think that the native grasses wear extremely
well. Most of them are deep-rooted perennials, so
they hold the country together. They need grazing to
compensate for the grazing of kangaroos, and I think
they would probably have a better filtration system
than conventional exotic species. There is now work
being done on your Kangaroo grasses [*Themeda
spp.*] to hybridise them so that they provide the same
nutritional value and house the same insects as exotic grasses. You won't get farmers to retain them on
their properties while there's little nutritional value in
them for stock—and that's that—so they've got to do
some work on the natives. They are also realising that
your deep-rooted perennials are really your quickest

way to go in salt prevention, because they transpire so much water. I mean, why not have your pastures working for you?

I don't think there will ever be salinity problems on 'Lanark', because dad faced up to the problem so many years ago. He probably didn't realise that what he was going to be doing by planting trees was preventing salinity. There's salinity 2 kilometres from us on the Smokey River—it's terrible. But dad's 'bowled the turkey before it's crossed the road'. It can't catch us now—the salt.

It's all about caring for the land, isn't it? That's why the trees are so important. In another twenty-five years time, I'll probably be thinking, 'You were right, you old bastard dad, putting those trees there. You were spot on'.

And there will be children climbing in them. You know that you're supposed to take out anything that is lying over in your agroforestry because it affects the growth of neighbouring trees—anything that's got a twisted trunk or lying on the ground. Not dad, no—he leaves one in about every 400 hundred metres and puts a big fluorescent orange ribbon

The Racecourse, an agroforestry plot of Pinus radiata.

around them, at chain-saw level, so that I don't cut them down. And he writes on the end of the tag, 'climbing tree for grandchildren'.

From a letter from David Fenton, 12 May 1998

Well the third generation has finally arrived, William David Fenton, born 15-08-96.

. . . Production is on the up at 'Lanark', increasing carrying capacity by 2.5 sheep to the hectare over the past five years. It has become completely obvious over the past several years that we must increase production on the most fertile land available to us on 'Lanark'. This must be done in order to compensate for initial loss of production due to large areas of trees and wetlands being established. This is undertaken by doing regular soil tests and applying specific fertilizers to the productive country, staying well away from the more native type areas, and using strict grazing formats which have been developed over the past twenty years.

One thing more: you must only plant as many trees for timber production as you yourself can manage in a year. Don't forget that you still have a conventional type farm to run and until the forestry begins to produce an income we must not lose sight of that.

PADDY FENTON

Paddy and Bronnie Fenton run 'Vasey Farm', located on the sandy clay loam country in the Dundas Tablelands approximately 50 kilometres northwest of Hamilton. Paddy took over management of the property from his uncle in 1989; it was previously run by his grandparents. 'Vasey Farm' has 5500 sheep and 100 cattle grazing on 688 hectares. Its distinctive giant River Red Gums (*Eucalyptus camaldulensis*) are typical of this part of the Western District, although many of them are reaching maturity and being lost either to age or storm damage. Paddy and Bronnie are concerned that as the tree cover on 'Vasey Farm' slowly declines, salting problems become a threat. A fifty-year farm plan is a strategy to retain Red Gum numbers and increase overall tree cover, compensating for any loss of productive land with agroforestry.

Paddy

I caused a bit of a stir at the first Landcare meeting. There was a fella from up the road and he said, 'Oh, I'm not worried about planting bloody trees. I'm in this for growing stock and growing wool, pastures is what they should be researching—getting the pastures right'.

And yes, sure they should be researching the pastures, and they are getting the deep-rooted pastures right. But I said to him, 'That's a pretty selfish attitude'. And he pricked his ears and looked at me as if to say, 'Who's this young upstart?'

And I said, 'Sure, your pastures are going to control the salinity in the short term, but you've got to start doing something about planting your trees, because all these big trees are going to start dying at some stage. You've got your son here, and he's got kids. You're getting the most out of your farm while you're alive, but you're not doing anything about maintaining your farm for your grandchildren's children, and their children after that'.

And he just laughed at me. Laughed in my face, like the old conservative cocky. But he'll change, everyone's changing. Because if they don't start planting trees and doing that sort of thing, they will be left behind.

We have developed a fifty-year farm plan, that focuses on retaining the tree coverage to prevent salting and to create agroforestry plantations for future income. Instead of superannuation policies, we plant trees for agroforestry. There will be more trees for soil protection in the gullies, as well as in and around the agroforestry plots. That way, when the agroforestry is ready to be milled, you don't think, 'Oh my god! I've taken my agroforestry trees down now and doesn't it look terrible?'

*The denser tree cover on 'Vasey
Farm' distinguishes it from
surrounding properties.*

I want every paddock to have shelter on virtually all fence lines, as well as individual trees scattered throughout.[1] To me, it would be just perfect to have a farm where every paddock is enclosed with trees and their shelter and ecology are working together.

If there is ever anything we want to know about trees, dad will come straight up here to show or tell us. Dad's a big influence on me. Our knowledge of caring for the land has been part of our upbringing—a part of everyday life. But mum is the one I have to thank for giving us the opportunity to take on 'Vasey Farm' as it belonged to her parents—my grandparents. Yep, she has been a great help in getting us started—and grandad's love of the tall timber was a bonus, you can see that in the landscape. When the ring-barkers came through sixty years ago he made them leave a lot more trees than some other properties around here. If only they knew what future damage they were doing at the time. It shouldn't have been called ring-barking, it should have been called murder!

You look at the massive timber in one of those Red Gum trunks. Touching fingertip to fingertip, you can get five people around the trunk. If all those trees started dying at once, this land would turn into a

wasteland. There is no way that you could control the salting, there would just be ground water coming out all over the place. One of those old Red Gum trees would be pumping about 200 litres of salty underground water per day, and up to 500 litres per day in the spring. It could take more than 1000 six foot saplings to equal that. Some of our neighbours, and a few people only a couple of mile as the crow flies either side, have shocking salt problems.

Bronnie: 'My father was always big on planting trees. He wasn't planting them for agroforestry or salinity control, but for shelter and aesthetic value. I knew of their value for shelter. I was aware of rural salinity problems, but it didn't mean much to me when I was living and working in Melbourne. I never realised how important trees were until I became involved with the Fentons.'

We lose about two big Red Gums a year from either wind or lightning. If lightning strikes one of those big trees, it will more often than not kill it because it burns all the gasses out of the tree. The tree may look all right for a while but it will die. There's one up in the paddock where the top of the tree has blown apart, and it only has a skinny stick left standing. The only reason that I can tell it's been struck by lightning is because of the red bits of wood scattered 60 metres around the tree—you'd be dead if you were near it when it happened. Lightning hits and the tree explodes, shrapnel flying—pretty powerful stuff! I don't know if Bron's ever seen one fall down in front of her, but I watched a tree come down one day which probably had 20 tonnes or more of wood in it. Twenty tonnes of tree coming down. As you watch the tree fall in the wind you feel like getting a peg to hold the thing up, or to try and stop it half way because the tree is something so big and so beautiful. It's been there for so long, and it's just falling down because a gust of wind caught it the wrong way—it nearly brings tears to your eyes.

Today parent Red Gums are being fenced from stock for regeneration.

AMANDA AND CHARLIE FAIRBAIRN-CALVERT

Amanda and Charlie Fairbairn-Calvert own and run 'Banongill East' on the slopes of Mount Widderin, an extinct volcano 8 kilometres south of Skipton on the eastern edge of the Western District. The property has three land types: basalt stony rises, plains and swamp depressions. The fertile and well-drained soils make good grazing country and 12 000 sheep are grazed over approximately 2045 hectares. 'Banongill East' is affected by relentless winds and a large rabbit population that is protected by the rocky landscape. Apart from stands of River Red Gums at the edges of the swamps, the property had been totally cleared of trees due to fires in the early 1940s and subsequent grazing. Charlie Fairbairn-Calvert moved to the tree-less eastern out-paddocks of the home farm, 'Banon-gill', in 1975. He began to plant trees for shelter near the existing homestead, but it was not until he married Amanda Fenton that a farm plan was introduced. Amanda and Charlie aim for a diversified income

Photograph courtesy Charlie and Amanda Fairbairn-Calvert

through sheep grazing, agroforestry and aquaculture, all coexisting in an integrated landscape.

Charlie

The home farm 'Banongill' was always the home of the rabbit. They spread up and down the lava flows between here and Derrinallum. They are just so hard to get rid of. You can poison them in the fenced-off areas because you don't have stock in there. In the open country you can riddle them out and chase them down, but in the rocks it is so hard. We've had a full-time shooter who's been here for a bit over a month, and he's taken 3000 pair of rabbits off the place, 6000 rabbits. Of those, most had four or five little ones in them. That's a lot of rabbits, so they have to do some damage to the land—and they do. The rocks harbour them. We were all too slack ten years ago when the myxo[2] was working. It used to wipe out most of the rabbits, and every autumn they would get down to a fairly low number, so no one

Photograph courtesy Charlie and Amanda Fairbairn-Calvert

Photograph, Lindsay Stepanow

worried about them. All of a sudden they became immune to myxo, and the numbers exploded.

They also live in amongst the grass and trees, and they trim the trees off at the top of the guards. It is so obvious, the damage they do. We have to guard all trees. You can buy the trees quite cheaply, but it costs you well over a dollar to get the guard on it. So instead of spending a dollar on a guard you could have bought another tree and a half. Instead of planting 1000 trees, you could have planted 2500.

Amanda

We set ourselves a target of planting 3000 trees each year. So far we have been planting more than that— 25 000 trees over six years. Our young son Clive will obviously see a lot more trees. He will help improve the property over the next fifty years.

Being brought up in the environment at 'Lanark', which was so sheltered, here I really noticed the wind, the cold and the fact that you couldn't just look out and see trees and walk amongst them. Charlie started planting before we got married. He had a few plantations around the house and the shed, up the hill by the house and along the road. He certainly got far more motivated when I arrived on the scene. In the beginning, we didn't decide what trees to plant, we just had all the left-overs that dad kept throwing in our direction. We started planting together and we haven't looked back.

Charlie

There wasn't a great number of birds at all before the trees, but there's certainly a lot of different species now. There's a black falcon around at the moment, which is very unusual. He just hovers above you as you walk around the sheds. I don't know why the falcon's hanging around this end of the farm, but it certainly has a lot to do with the rabbits.

Amanda

Charlie thinks that a target of 10 per cent tree coverage for the farm is not beyond the realms of the imagination—as long as you have woodlots and other things that can make you a bit of money in the long term. On our property that would be 500 acres [about 200 hectares] of trees. We will do 5 per cent and let Clive do the other 5 per cent. We've got to leave something for him to do—we can't just hand it all to him.

Charlie

Ten per cent is a fair figure, but I can see a much bigger percentage if some of the woodlots are turned to

agroforestry. There is a farmer down the road trying to make any tree that he plants pay its way. It seems to be a fairly sensible way of increasing your income without cutting your sheep numbers along the way. They just seem to blend in well. They compliment each other, the trees and the sheep. The trees reduce wind stress on the animals and pastures, increasing both wool production and the pasture's resistance to drought.

In the future I don't see any reason why Clive shouldn't have a pruning saw in his hand when he goes out to feed the sheep. Why David already does!

From a letter from Charlie Fairbairn-Calvert, 8 May 1998

We have had a big drop in rabbit numbers due to the Calicivirus,[3] and Myxo working again. As a landcare group we are also doing a lot ($80 000) of Harbours ripping[4] to stop them coming back.

The very low wool prices for the last six years has seen us diversify our income even more. We now sell clover-hay, fat lambs, grain and have paid shooting of rabbits.

The future has arrived, we have Blue Wrens in the garden. I have pruning saws under the seat of my ute and Clive (six) and Anna (four) are helping to

plant and prune the trees. Let's hope we are around
to cut them down.

Notes

[1] Mature River Red Gums (17 per hectare) in a paddock at
'Vasey Farm' show that at 1.5 metres above ground, the
trees reduce wind speed to 50–60 per cent of that in an
adjacent open paddock (Knight 1989).
[2] Myxomatosis, a highly infectious viral disease of rabbits, introduced to reduce rabbit populations in agricultural areas.
[3] Rabbit calicivirus disease is a naturally occurring disease in
European rabbits. As feral and pet rabbits in Australia are
derived from European rabbits, trials to test the susceptibil-
ity of Australian rabbits to calicivirus as a biological control
agent were undertaken on Wardang Island off the South
Australian coast. The virus's escape to the mainland during the trials in 1995 was followed by official release programs to eradicate rabbit populations around the country.
[4] The deep ripping of the soil with a tractor mounted
implement to destroy the rabbit burrows.

I come from Lake Condah

Johnnie Lovett

In a spiritual sense, I get a lot of growth from being in the Lake Condah area. I feel that I know where the song began, but I don't know the song.

My responsibility as a custodian is to ensure that the culture of the area is spoken about with truth, sincerity and feeling. Aboriginal history in this area is 29 000 years. I believe we have been here since the beginning.

The Eagle Aboriginal Tour aims to create understanding. I believe we can't build bridges between cultures in the future if we don't look at the past and understand what happened. People should feel the cultural spiritualism that's connected to the area. To view the fish traps and stone house sites.

What I'd like people to understand when they go to Lake Condah is that they're in an area where traditional people have walked, talked, danced, and told stories; to feel the country.

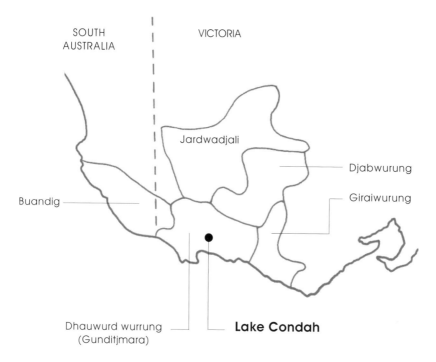

Indigenous language groups of the south-east coastal region surrounding Lake Condah and Gunditjmara country

Lake Condah lies 15 kilometres northeast of Heywood, in Victoria's Western District. It was formed 27 000 years ago when Mount Eccles erupted, creating the Tyrendarra lava flow, which gave birth to the basalt Stony Rises, lakes, swamps and creeks throughout the region. The lake was part of the Dhauwurd wurrung[1] people's territory, the names of whose clans derived from features of their country: *kerup* (water), *bum* (mountain), *direk* (swamp) and *gulger* (river) (Gould n.d.). The clans surrounding Lake Condah itself became known as Kerrupjmara—the water people.

Life thrived in the extensive seasonal wetlands throughout the Western District after the autumn and winter rains, and provided an important food source for the local Koori people.[2] Kerrupjmara took advantage of the lake's offerings and built fish traps using the local basalt stone. They engineered V-shaped channels to divert water into weirs through eel baskets woven from bark strips or plaited rushes, creating an effective harvesting system. The system took advantage of topography, with changing water levels determining the use of traps at certain heights. This technology, together with the climate of Victoria, determined a semi-settled society, leading to the construction of semi-circular shelters. Local basalt, used for low walls, was covered with rushes and bark supported on wooden frames (Flood 1990, 1995).

Whalers and sealers began to work the Portland Bay area south of Lake Condah in 1810, bringing seasonal European contact to the local clans. The coastal Dhauwurd wurrung clans were subjected to violence and disease. In 1834 the permanent white settlement of the Port Phillip colony (Victoria) was led by Edward Henty who crossed from Van Diemen's Land (Tasmania) and settled in the area that was to become Portland. Conflict between the local Kooris and the new settlers over land use developed as sheep were introduced to graze what was to become part of the continent's prime wool growing region. Pastoralists opened vast tracts of land, waterholes and rivers for their stock, occupying many traditional hunting and foraging areas, meeting places and sacred sites. Increasing stress on traditional land uses escalated until the 1840s when Kooris launched attacks on settlers. This war of resistance, known as the

Eumerella War, included attacks on properties, theft
of horses, clothing and food, and spearing of cattle
and sheep for meat as local food resources dimin-
ished (Clarke n.d.). Retaliation against these attacks
included massacres, shootings and poisoning. Within
fifty years of white settlement, almost the entire
'full-blood' Dhauwurd wurrung population had been
destroyed (Cole 1984).

In 1858 a state government committee investi-
gating the problems facing the remaining Koori people recommended establishing reserves on traditional
lands, to be run by missionaries. The Central Board
for the Protection of Aborigines was formed to implement the committee's proposals. In 1869 the Central
Board became the Board for the Protection of Aborigines, a statutory authority, and by 1874 mission and
government stations controlled over 24 000 hectares
of land.

A mission was established at Lake Condah in
1867 by the Church of England Mission for Aborigines (Cole 1984). It provided a degree of safety from
the violence, but at the expense of indigenous culture, which was largely lost through the teaching of
English language and Western lifestyle and religion.
The 1880s marked the height of mission activity before the Aborigines Protection Law Amendment Act
in 1886 prohibited any 'part-Aborigine' or 'half-caste'
from living on stations. The Act was controlled by the
Board, whose powers determined how and where
Aboriginal people lived, preventing any degree of in-
dependence. The population of the mission dropped
from 117 to 20 and, although the mission fell into
disrepair, Kerrupjmara lived there until the 1950s. In
1951 the government handed over the land of the
Lake Condah reserve to the Soldier Settlement Commission. Except for the land occupied by the mission
buildings, the cemetery and its access, it was then
divided into farmlets for returned soldiers. In 1984 a
struggle for land rights resulted in 53 hectares being
handed back to the Kerrupjmara (Clark n.d.).

Since the land was handed back to its traditional
owners, assessments determining the significance of
the Lake Condah site have been carried out. During
1989 and 1990 the Victorian Archaeological Survey
(VAS) recorded 87 shelter sites, 78 fish trap sites and
4 artefact scatters. The Lake Condah area and similar

sites throughout the Stony Rises region represent over 200 square kilometres of Aboriginal stone structures and related sites (VAS 1992). A review of world archaeological records indicates that examples of stone shelter sites and fish traps of hunter-gatherer societies are rare, and in 1992 VAS assessed the Lake Condah site for world heritage listing.

Remnants of Lake Condah
Mission

JOHNNIE Lovett is a descendant of the Kerrupjmara and was born and raised in the Lake Condah area. After Lake Condah was returned to its people, Johnnie initiated the Eagle Aboriginal Tour Company in 1992, sharing with visitors the area's significant sites, including the remains of stone shelters and fish traps that date local culture back 29 000 years, and areas where his people were massacred. In 1993 the company was given a five-year contract by tribal elders to run the tours.

Johnnie sees tourism as a way to forge greater understanding of local Koori culture and move Aboriginal people away from welfare dependence. Although much of the local culture has been lost, Johnnie displays a strong commitment to his home-land and a connection with the spirit of his people.

Johnnie

The Dreaming[3] is a part of where you belong. I come from Lake Condah and it's never been in stories handed down to me that I've come from anywhere else. I think every indigenous group throughout Australia has the same thing. I believe that there have never been any stories handed down to them from the Dreaming that they've come from anywhere other than the country of their people. Even those who have been adopted by white families and have grown up in white society trace their blood lines back to certain areas, certain communities, certain families.

In a spiritual sense, I get a lot of regrowth from being in the Lake Condah area. I get a lot of strength from being here because I can feel the spirit of my people. If I'm away from home, once I come back into the area and I see the Grampians mountain range, I'm coming home, ya know? I can understand that in traditional times one would come into the Kerrupjmara homelands, and sing their song right through it. I feel that I know where the song began, but I don't know the song.

On me travels to meetings in different areas, I drive through Victoria and I always see something that I can firmly relate to my Aboriginality, or to Aboriginal culture, and I'm pretty glad I've got that. I can drive through an area and have a direct link in my thinking with what I see, whether it be a Bottlebrush [*Callistemon spp.*] flower that my people used to drink water through as a sweetener, the Redgum that they carved their spears and boomerangs from, or the ovens where food was covered in reeds and cooked by the heat of basalt rock placed in the fire. I can see where my people had been in times gone. Most people wouldn't relate that culturally, like I can and do.

My schooling was very, very limited. I only went to half way through second year of high school. I left for reasons of racism 'cause I was the only Aborigine who attended secondary school at that time, in a school of about three hundred. That was in Heywood, and at that time we lived at a little place called Greenvale Sunday Creek about 4 mile out of Heywood. To the wider community that lived in Heywood, we were the fringe-dwellers.[4] I use to walk out of the bush every morning and catch a bus into a town full of white people, a school full of white people and white teachers. In those classrooms I was not given the recognition of being an indigenous person of this country. I was told that Captain Cook discovered Australia and in the afternoon I'd catch the bus back home to my people.

My father and mother married at Lake Condah Mission in 1926. Dad served in two world wars. He'd been to the First World War and he came back and took five of his brothers to the Second World War. After they came back from the war on leave, they were refused a beer at Lake Condah Pub. My father and two of his brothers and my uncle were refused a beer in full uniform because they was Aboriginal. Then when the mission closed down, it was cut up into soldier settlement blocks after the Second World War but my father didn't receive any block—him or his five brothers. When he'd finished his service for this country, he was given nothin'. Ironically my father, Herbert Staley, was named after the last missionary of Lake Condah, the Reverend Staley.

My father was the organist at the church. The church played a very big part in the lives of my people, and it did to me as a kid. Sunday School was always held at the church, and of course going back to the 1800s, when the mission was in full swing, the church really played a dominating role over the

people. They had to attend in the morning and in the afternoon—Church of England. That was the law of the missionary of the day, and it stopped my people from being in their traditional areas; stopped them from going out and teaching the kids the song and the dance and the spiritualism. The missionaries then handed out tea, flour and sugar which stopped people from practising their hunting and gathering techniques because they had no excuses for being off the mission. It was a very, very supervised exercise that was meant to cut the culture. It took away the language, the song and dance, the law, the spiritualism and the religion. I see that missions were only built to serve the purpose of opening up the land for non-Aboriginal people.

I mean the missionaries came here with the belief that we were heathens and children by nature, y'know? But no children could have survived like we've survived, and today it stands as gospel, that we have survived over thousands of years. And when you take into consideration what's been thrown against us, the acts of genocide: poisoning the water and the arsenic in the flour, the introduction of syphilis and gonorrhoea, even the fact that small pox was devastating, the common cold was devastating; and we have still survived.

The Lake Condah area is very, very rich with massacres. There was a property just west of where I'm living now, which was called Wedges Property, and they had a Gatling gun on that property to deal with Aboriginal people. Malcolm Fraser, the ex-prime minister of this country, lives not far from here, and near his property there is a place called Penongatong Reservoir, or 'the fighting water-holes' as we know it. People were massacred and dumped. Their bones were dumped in shallow graves on the banks or mixed with sheep bones and dumped in the reservoir itself so others couldn't identify the difference. But we were a very strong group of people, to the extent that it was the magistrate of the Glenelg area who brought in Captain Dana and the Native Police Corps[5] from Melbourne for the duration of what it took to quell the disturbances created by us, which actually meant that they massacred everyone. The Nillangunditj people were wiped out. They were massacred in Gorrie Swamp. I suppose it would read a little bit different in

'In a spiritual sense, I get a lot of regrowth from being in the Lake Condah area. I get a lot of strength from being here because I can feel the spirit of my people.'

official materials, but my grandmother told stories of massacres happening when she was about nine or ten. She was a little girl and hid in a cave on the other side of Darlots Creek when they happened.

Today, I'm one of the seven male custodians of the Lake Condah area, in a sense that there are seven of us that still belong to the Lovett, Saunders or Alberts families which are the main families from the area. So I'm one of the seven males of Condah, and its pretty sickening to think that at 46 years old you can be one of the last people that is related directly to the area. I pass on knowledge of the area to my children and relations and they will have a choice of whether or not they want to continue; to tell the history of Condah and relate it to Aboriginal groups and non-Aboriginal groups throughout the area, or whoever comes around at the time.

My responsibility as a custodian is to ensure that the culture of the area is spoken about with truth, sincerity and feeling, and what people experience at Lake Condah is the truth. They learn about the stone fish traps and the stone house sites that date back to a time when the traditional lifestyle at Lake Condah was very much in full swing. They can see that the stone houses made by my people were part of their living environment and the fish traps were part of a

life cycle. It shows how well my people understood the surroundings in which they lived because they sustained a steady food supply over a long period of time, and today you can still go down and catch a fish in the traps. People may not fully understand that when they talk about ancient Egypt they are talking about 7000 years of history. Aboriginal history in this area is 29 000 years, and that's by white man's record. I believe we have been here since the beginning. We were the beginning.

Today you see a lot of young kids running around here, or me grandchildren, and me nephews and nieces, and they need to know, they need to be given the opportunity to feel the way I feel about the area. And I need to be able to take them there and teach them, regardless of whether later on in life they wish to continue the teaching or not. But they should be given the opportunity to be able to say that they know the area. And I think that that's important because it strengthens their Aboriginality.

Canals were excavated in basalt rock, and were obstructed with free-standing stone walls with openings for nets or woven fish traps.

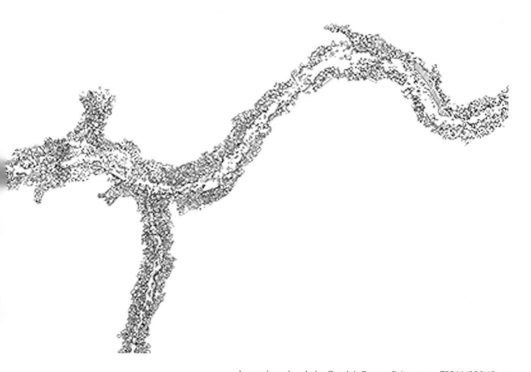

Image based on Lake Condah Survey fishtrap no. 72211/086 (Coutts, Frank & Hughes 1978).

But the site is not only important for teaching the Koori people. I want to promote Aboriginal culture so that racism is not as intense as it was when I was growing up. When I was a kid my Aboriginality was strong and there were always questions in my mind as to why racism was so strong against Aboriginal people. The Eagle Aboriginal Tour Company aims to create understanding. It increases the number of people visiting the Lake Condah site and brings more publicity to Aboriginal culture. I've seen a lot of evidence that the tours are bridging the cultural gap. For instance in 1987 I had people come in from different walks of life. There were professional people such as teachers, university educators and solicitors. Amongst these people was a lady of Jewish origins whose mother was caught up in the Holocaust. She carried the star of David in her bag, but she couldn't explain why she happened to bring it to Lake Condah until we sat and talked. We talked about atrocities caused to peoples and the understanding that we need so it doesn't happen again. This lady had something in her life that she couldn't find the

answers to, but through the exchange we had that night she found something in Aboriginal culture. That lady cried.

As a whole, today's wider society does not have a full understanding of our culture, although I must say that the education departments are now trying to insert Aboriginal studies within the curriculum. It should have happened in primary school, in the beginning, you know, so that you grew up with the knowledge of Aboriginal people. Today, I feel that before non-Aboriginal people can take the line of telling you that they are Australian they have to understand Aboriginal culture, because if they want to be part of this country, we are the country. We are the country. The land is not ten thousand sheep on acreages of land. It's the spiritual and religious part of being what we are, as black people.

So the significance of Lake Condah for non-Aboriginal people, is to experience a sense of history. I believe we can't build bridges between cultures in the future if we don't look at the past and understand what happened. And I think the past is a good foun-

Aboriginal history in this area is 29 000 years.

dation block for any building that's going to stand in the future. People have to recognise that there was massacres that happened at Lake Condah. People should feel the cultural spiritualism that's connected to that area. To view the fish traps and stone house sites and say, 'Well, these weren't just a group of people who walked around with the boomerang and spear and hunted kangaroos and emus and that was their lot in life', because the technology that we had to fish was something that, I believe, no other race of people would have done.

What I'd like people to understand when they go to Lake Condah is that they're in an area where traditional people have walked, talked, danced, and told stories; to feel the country. Feel the country. Sit down and put your hand on a stone, and feel the warmth of the sun in that stone. Put your hand against that tree and feel it swaying, because it too is a living thing. That's the spiritualism of being an Aboriginal person —understand my part of being a black. And if people can do that, I think that's a bit of a start. That's a bit of a start for building bridges in the future. That's what I want to get through to people, how we feel about the land.

So I think there needs to be a full knowledge and understanding of Aboriginal people to the wider community. But I say that with a smile on me face simply because how do you teach 140 000 years of culture, which is the figure now bandied around

between government departments and Aboriginal groups, in the period of one's lifetime? How do you understand 140 000 years of song, dance, language, spiritualism, religion, and technology?

I guess some of the things that people don't fully realise is that there is so much missing in Aboriginal people's lifestyles. Government policy, the massacres, the missions, they resulted in a loss of traditional life, a loss of language and passing knowledge, a loss of culture. Not knowing my traditional name or being able to speak my language is something I was born with, and something I'll die with. They are the missing links to my Aboriginality, and I feel very strong about those things. There's been a lot lost, but I can be thankful for the traditional areas that I have an affinity with. I'm thankful for those.

We as Victorians certainly do have a culture within ourselves and within our communities. I feel put out when I see a lot of shops of central desert and western desert art and Aboriginal artefacts being publicised in the state of Victoria. To me it's saying there is nothing else here, y'know? I think that it helps the wider community think that either there are no Aboriginal people in Victoria, or Aboriginal people in Victoria don't have a culture.

The survival of our culture is going to be in the strength of the young ones running around here today, it's going to be their strength to build a future on their Aboriginality. And learn to understand it and accept it for what it is, and not become complacent with the fact that they are Aborigines, not to feel that they are inferior in any way to other people who aren't Aborigines. I think we are gonna survive, because our kids are feeling a stronger sense of their Aboriginality, and they aren't ashamed of it today. Some years ago people were ashamed of what they were, but we have strengthened.

Sometimes it's very hard when you try and emphasise your Aboriginality as a part of you, wanting to keep your identity as strong as you can. And then you have people say 'Yeah, I know how you feel, yeah, we're all the same'. We are not the same, we're as different as chalk and cheese and I think people have to realise that. I mean look around you, you can see this is a very conservative area because of the land and the farms. They have never heard their mother crying because there is nothing to feed them the next morning. They have never had to walk like my mother did, with a sugar bag on her back, whether it was rain, hail or snow, to go into town and walk all the way back again with tucker. And they say to me, 'We know how you feel about the land, because the land's been in our family for six generations'. I always say, 'Well the land's been in my family for two hundred generations'.

Today I think things are better because the wider community are a different breed of people. We have people who are curious, we have non-Aboriginal children in schools who are not afraid to ask questions, and who are pretty sensible in their thinking, and I think that their attitude will certainly cut down the level of racism. I think there is a greater understanding: I had some thirty-five non-Aboriginal people on a tour in Lake Condah on the day of the bicentenary celebrations of this country. We was all standing around the flag pole and we had the Koori flag lowered at half mast. I was giving a talk and a white woman said to me, 'John, look!'

And I said 'What?'

She said, 'Look up there, your totem!'[6]

Two wedge-tailed eagles flew out over the mission, were joined by a white eagle and flew over the flag pole once, and then off into the stones and over the lake. I have thirty-five white people that will tell you that, and if that is not the sign of strength, or if that's not an omen of culture, then I'm not Aboriginal as far as I'm concerned.

I'm a regional ATSIC [Aboriginal and Torres Strait Islander Commission] councillor with the Tumbukka Regional Council which involves a lot of communities down western Victoria right through to Sunraysia area. When you sit down and look at the dealings of government departments within those communities and the funding that's been generated into those communities, I can't help but liken it to the ration days of the 1800s when the missionaries handed out tea, sugar and flour; when they took away the hunting rights of Aboriginal people, along with the law, song, dance, language, traditional names and so forth. I can't help but liken today's handing out of dollars and cents by the government as another form of trying to keep the mission mentality in the Aboriginal people, keep a dependence in Aboriginal people so they can be controlled.

I use to get angry. I grew up angry, and only since I got to my early thirties that the anger started to leave me. I started to look at things differently. One of me biggest dreams is acceptance of Aboriginal people as equal, because for too long we've been on the flip side of the record and always have been treated as such. Now being such a multicultural country, we see a lot of cultures within the structures of present-day communities in Australia. I don't really have a problem with that. The problem I do have is that I never want to become part of a mainstream multicultural, ethnic group. I respect other people's culture for what it is, but I want to maintain my identity as an indigenous person of this country. I have high hopes that one day a lot of families will come back and will regenerate the cultural hunger within our own people for them to want to learn, to want to come home and carry on where people like me left off. That's a dream of mine.

My vision is to employ twenty Aboriginal people in different areas, to restore some of the old houses on the mission, to restore some of the fish traps and old stone house sites and to have a steady flow of people through the area. The Eagle Aboriginal Tour Company is about educating, and knowing that out of a group of thirty or forty people, ten or fifteen will go away with a different point of view is a personal

gain. They are Australians, or part of this country, so in a sense the culture also belongs to them. I hope that when my time comes to go, people will remember me for being a part of the culture that they have learned.

As a young person growing up in a mission, I missed out on things that dealt with me as an Aboriginal person, so the only time I'll really be free to join my people is when I die. It will be the release of the spirit for me. It will be the freedom to be wherever I want to be, to be a part of whatever I want to be a part of, and that's a very strong thing in my mind. I will be cremated and my ashes will be spread in Lake Condah itself, and I firmly believe that I'll come back as the tree, the rivers, the wind. I'll be a part of everything.

Notes

[1] The Dhauwurd wurrung people of over 56 clans are more commonly known as the Gunditjmara. Their country extended from the Glenelg River in the west to the Eumeralla River in the east, and from the southern coast to Mount Napier and Hotspur in the north (Clark n.d., Cole 1984).

[2] 'Koori' means 'man' or 'people' in several languages of southeastern Australia. Since the 1960s, it has been commonly used as a general term for indigenous people of New South Wales and Victoria.

[3] Dreaming 'is that timeless epoch of creativity that gave form to the diversity of life, set in motion nature's cycles, and left its enduring imprint on the earth's crust—all species including kindred humans, were subtly entwined within a transcendent web of meaning that renders eternally sacred, the process, places, and personages of the natural world (Knudtson & Suzuki 1992, p. 39).

[4] Fringe-dwellers are indigenous people living in self-constructed shacks on the edges of towns throughout Australia. Fringe-camps 'have developed as part of how Aborigines have attempted to control the effects of the increased power and involvement of various social welfare agencies in their everyday life . . . to assert that fringe-dwellers are simply detribalized people is to transform an important political act into a symptom of individual affliction and group decay' (Collmann 1988, p. 3).

[5] H. E. P. Dana was Commandant of Native Police, a corps of Aboriginal police established by Governor Charles La Trobe. A detachment of the corps was stationed in the Western District from 1842 to 1849 (Fels 1988).

[6] 'Totem' derives from a North American Indian dialect and means he/she/it is a relative of mine. Totems are essentially symbols of relationships: clan totems symbolise the relationship of clan members to each other, to their ancestors and the past, and to particular places or sites; conception totems symbolise the relationship between people and their places of conception, as well as the individual and the ancestral spirit world; and sex totems symbolise the relationship between people of the same sex and their differentness from the opposite sex. These symbols play a key role in ceremony as they deal with relations between people, to the landscape and the ancestral past (Peterson 1994).

Epilogue

Jim

While the first ideas for this book were conceived in 1991, unconscious research had begun a decade earlier—after I came from Iowa State University to begin and head the new undergraduate landscape architecture course at the Royal Melbourne Institute of Technology.

Before arriving in Australia, my interest in broad-scale land management issues had led to directing the Land Use Analysis Laboratory at Iowa State University. I always attempted in my teaching to have my students see site-specific projects through an understanding of broader issues. I led a student team to produce a photographic exhibition showing the land patterns of Iowa from the air and ground, which captured the quilt-like beauty of this rural landscape. These experiences always made me feel as if my life had a urban/rural duality, because my early childhood was spent exploring and playing in Flushing Meadows, just outside New York City. This duality would play out again in Australia. Although I moved to Melbourne, valuing the rural landscape and wanting an introduction to the lands of my new home, I welcomed every opportunity to visit rural and remote areas, to gain at least an elementary understanding of the issues facing landscapes and peoples beyond the urban fringes.

I felt that my new awareness of these Australian landscapes might both help my students and extend the boundaries of my urban profession. As many of my students were from urban backgrounds with limited personal experiences of broad landscape issues, I set out to secure links with rural towns and pastoralists so that students would come face to face with the issues and meet the people affected by them. Together, we experienced first-hand the beauty, spirit and people of rural and remote Australia.

So the RMIT OutReach Australia (ORA) Program, first called landscape windows, was informally established in 1983 as one of the design studio options for students. The program has been critical in advancing both student and staff experiences in rural and remote Australia. A variety of projects has included land management strategies, townscape and community planning, landcare and tree planting programs, development of structures for landscape-related employment programs, and identification of directions for alternative settlement planning strategies with indigenous communities that promote the landscape as a critical focus for many issues associated with Euro-centric settlement models, especially poor health. Through the program, links with some of the people in this book were established.

My original concept for the book centred on my attitudes about landscape and design, not rural and remote Australians. As a landscape architect I had been influenced by prominent ecologists, landscape architects, architects and planners including Jack McCormick, Ian McHarg, Roberto Burle Marx, Aldo Guirgola, Lou Kahn and Edmund Bacon—my teachers and mentors at the University of Pennsylvania. But now, rather than write about my own experiences, I felt that there were more important messages to be shared through the voices of others. I invited Phin to join me in this first project of our working relationship.

Phin

When I started the landscape architecture course in 1987, the two primary focuses of the ORA program were conceptualising regional landscape schemes for the wool growing properties of the Western District in Victoria, and developing strategic planning regimes for the arid town of Coober Pedy in South Australia. In 1989 I visited Broome with sixteen fellow students to review water issues and design a water-retention proposal for the developing peninsula. For most of us, this was our first contact with indigenous people and their culture. During my final year I worked with an Aboriginal community in the Kimberley and developed a landscape improvement plan under Jim's supervision. This project continued with other students, and after I graduated it became the platform for a working relationship with Jim to better understand the potential role of landscape architecture in working with rural and remote indigenous communities.

During 1992 we visited the people we had met through ORA projects and met others with similar attitudes about country and culture. We travelled an imaginary line between the Kimberley in Western Australia and the Western District in Victoria, recording the voices of people telling of their relationship to landscape, whether cultural, historical, spiritual or economic.

We have been encouraged by the responses of people who have heard or read these stories, whose quality is timeless. The stories demonstrate the profound impact that practical responses can have in solving cultural, environmental and economic problems, not only in rural and remote areas but also in the urban environment. We have been inspired by these people and by the landscapes that they have established or to which they continue to belong.

Australia is at a cross-road, and many of our economic choices hinge on culture and environment. In order to resolve the disharmony resulting from Australia's recent history—settlement of the continent by Europeans—we must recognise the vital connection between indigenous Australians and their countries. This recognition is essential not only for reconciliation but also for validating the importance of indigenous knowledge about the land and its potential to help establish sustainable land use practices. We must acknowledge, of course, that indigenous people living in rural and remote areas have responsibility for conservation throughout their countries, as do nonindigenous people for their various land uses throughout the continent. Our willingness to establish sustainable farming, mining and development practices is critical not only to the health of the Australian landscape but also to the health of the nation.

The stories in this book are evidence of knowledge held by individuals throughout Australia about country and culture. This wealth of local knowledge, gained through personal experience and trial and error, is often missing from debates about strategies for the future. It is an important resource for a nation that is beginning to see the need for more appropriate— and productive - land practices. We have been impressed by the knowledge we have encountered while travelling the country, and we hope that the messages of these voices will be heard, and perpetuated through story-telling.

Listen to the people
Listen to the land
What they're all saying
Can we understand

Acknowledgements

We wish to especially thank the storytellers who were so open and supportive in the production of this book. Without their experiences and commitment to landscape this project could not have come to life. A special thanks goes to Archie Roach who wrote the introductory song.

We also thank those at the Royal Melbourne Institute of Technology (RMIT University) who provided support, including the provision of facilities by the Centre for Design, the Faculty of Environmental Design and Construction, the Department of Planning, Policy and Landscape, and the Landscape Architecture course. A special thanks goes to Dr Michael Berry, past Head of the Department of Planning, Policy and Landscape; Professor Leon van Schaik, Dean of the Faculty of the Constructed Environment; Professor Chris Ryan, Director, and Henry Okraglik, past Associate Director, of the Centre for Design, and Dr David Knowles Pro-Vice-Chancellor for their personal support and enthusiasm. The Audio Visual Laboratory in the faculty assisted in photographic needs, the Communication Services Unit Print Centre assisted in tape transcriptions, as did Rochelle Ruddock, Simon Ridgeway, Nick Loschiavo and Yarra Hardjadibrata. Richard Thomas was a time saver when it came to help in word processing. Thanks must also go to the Landscape Studies class of 1992 who discussed and reviewed initial ideas on the book structure. David Jones provided studio space and loaned us his computer facilities, as did John Burgess during the main production phase.

Steve English, Jeanné Browne, Matt Campbell and David Hay were valuable OutReach Liaisons, cutting distance when we needed additional information but were in Melbourne. The following people provided information during discussions on our trips: Dr Rod Bird, Research Scientist, and Don Jowett, Landcare Officer of the Pastoral and Veterinary Institute, Department of Agriculture in Hamilton; Roslyn and Clive Philpott; Cath Krilly, Janet Skewes, Grant Drummond, and Neville Hyatt in

Coober Pedy; members of CSIRO in Alice Springs; Stanley Nangala and other members of the Department of Aboriginal Affairs in Perth; Teddo Cox and Sandy Paddy at Goolarabooloo Community and Ladjardarr Bay Aboriginal Corporation. Ansett Airlines in Melbourne provided airfare support.

In the area of photography, Frans Hoogland, Jo Henry, Bob Purvis, John and Cicely Fenton, and Charlie and Amanda Fairbairn-Calvert provided additional photographs to our own. Professional photo-graphy was contributed by Ian Oswald Jacobs, Lindsay Stepanow, whose aerial photographs were supplied through Thomson Hay & Associates, and Kay Johnston. Acknowledgement must go to the work of Dr David Horton who provided a valuable resource in his book *The Encyclopaedia of Aboriginal Australia*. The accompanying map of Aboriginal Australia depicting the continent's language groups was used to show locations of language groups mentioned in the stories.

John and Cicely Fenton, Charlie and Amanda Fairbairn-Calvert, David Fenton, Paddy and Bronnie Fenton, Yarny at Oomona Lodge, Laurie and Penny Cox, Veistures Ceilens and Kay Johnston, Nick Eastman and Chris Bird all provided accommodation when we were not camping.

A very special thanks goes to Curtis Sinatra who read, critiqued, discussed, word processed, transcribed, chased information, assisted in computer malfunctions and put up with us for the duration of the project. Jim thanks Richard Greenwood, past RMIT student, who introduced him to the outback.

Final thanks go to Peter Corrigan, Architect, who initiated contact with Melbourne University Press; Teresa Pitt, Commissioning Editor at Melbourne University Press, who was responsible for commissioning publication, and the production team who were enthusiastic throughout the publishing process, especially Gabby Lhuede, Production Controller, and Jean Dunn, Senior Editor.

References

Benterrak, K. Muecke, S. & Roe, P. 1983. *Reading the Country*, Fremantle Arts Centre Press, Fremantle WA.

Bird, R. n.d. 'Trees in Western Victoria: An Historical Perspective', *Trees and Natural Resources*, vol. 28, no. 1, pp. 8–11.

Bird, P. R., Bicknell, D., Bulman, P. A. *et al.* 1992. 'The Role of Shelter for Protecting Soils, Plants and Livestock', *Agroforestry Systems*, Kluwer Academic Publishers, no. 20, pp. 59–86.

Bonney, N. 1994. 'Lanark' Farm Forestry: An Historic Overview', in R. Ruddock (comp.), *'Lanark' Farm Forestry*, 2nd edn., privately published, Melbourne.

Bride, Thomas Francis 1983. *Letters from the Victorian Pioneers*, ed. C. E. Sayers, Lloyd O'Neil, Melbourne, pp. 168–9 (letter from J. G. Robertson, 1854).

Clark, I. D. n.d. *People of the Lake: The Story of Lake Condah Mission*, Koorie Tourism Unit, Victorian Tourism Commission, Melbourne.

Cole, K. 1984. *The Lake Condah Aboriginal Mission*, Keith Cole Publications, Melbourne.

Collmann, J. 1988. *Fringe-Dwellers and Welfare: The Aboriginal Response to Bureaucracy*, University of Queensland Press, Brisbane.

Coober Pedy Progress and Miners Association 1983. Preliminary Study for the Upgrading of the Coober Pedy Water Supply (unpublished).

Commonwealth Scientific Industrial Research Organisation (CSIRO) n.d. *Managing Rangelands*, Information sheet.

CSIRO 1991a. *History and Ecology of the Pastoral Rangelands*, Information leaflet, Centre for Arid Zone Research, CSIRO Division of Wildlife and Ecology, Alice Springs.

CSIRO 1991b. *Sustainable Pastoral Management and Land Restoration*, Information leaflet, Centre for Arid Zone Research, CSIRO Division of Wildlife and Ecology, Alice Springs.

CSIRO 1992. *Rangelands Assessments, Monitoring and Forecasting*, Information leaflet, Centre for Arid Zone Research, CSIRO Division of Wildlife and Ecology, Alice Springs.

Coutts, P. J. F., Frank, K. R. & Hughes, P. 1978. *Aboriginal Engineers of the Western District Victoria,* Records of the Victorian Archaeological Survey, no. 7, Victorian Department of Conservation, Melbourne.

Crilly, K. 1990. *The Discovery of Coober Pedy*, a 75th jubilee publication for the Coober Pedy Historical Society Inc., Coober Pedy.

Fels, M. 1988. *Good Men and True: The Aboriginal Police at the Port Phillip District 1837–1853*, Melbourne University Press, Melbourne.

Flood, J. 1990. *The Riches of Ancient Australia: An Indispensable Guide for Exploring Prehistoric Australia*, University of Queensland Press, Brisbane.

Flood, J. 1995. *Archaeology of the Dreamtime: The Story of Prehistoric Australia and its People*, rev. edn, Angus & Robertson, Sydney.

Goolarabooloo Association Inc. 1992. The Land of the Waterbank Pastoral Station: A submission revealing a unique opportunity to develop the land as an invaluable asset for the benefit of all (draft submission).

Gould, M. n.d. *Lake Condah Mission Station: A Report on the Existing Condition and History*, (with a survey of archaeological sites by Anne Bickford), Victorian Archaeological Survey, Melbourne.

Hall, N., Boden, R. W., Christian, C. S. *et al.* 1972. *The Use of Trees and Shrubs in the Dry Country of Australia*, Australian Government Publishing Service, Canberra, quoted in Bird, P. R. *et al.* 1992, 'The Role of Shelter for Protecting Soils, Plants and Livestock', *Agroforestry Systems*, Kluwer Academic Publishers, no. 20, pp. 59–86.

Hansen, K. C. and Hansen, L. E. 1992. *Pintupi/Luritja Dictionary* 3rd edn, Institute for Aboriginal Development, Alice Springs.

Hay, D. 1993. Nomination for Stihl Australian Forest Grower Victorian Tree Farmer Award: John and Cicely Fenton, 'Lanark', Branxholme (courtesy John and Cicely Fenton).

Horton, D. 1994. *The Encyclopaedia of Aboriginal Australia,* Aboriginal Studies Press for the Australian Institute of Aboriginal and Torres Strait Islander Studies, Canberra.

Institute of Foresters of Australia 1989. Trees: Their Key Role in Rural Land Management, Submission to the House of Representatives Committee of Inquiry into Land Degradation in Australia, quoted in Bird, P. R. 1992, 'The Role of Shelter for Protecting Soils, Plants and Livestock', *Agroforestry Systems*, Kluwer Academic Publishers, no. 20, pp. 59–86.

Isaacs, J. 1987. *Bush Food, Aboriginal Food and Herbal Medicine*, Weldon Publishers, Sydney.

Kenneally, K. F., Edinger, D. C. & Willing, T. 1996. *Broome and Beyond: Plants and People of the Dampier Peninsula, Kimberley, Western Australia,* Western Australian Department of Conservation and Land Management, Perth.

Knight, M. 1989. Preliminary Shelterbelt Wind-test Results in Mature Random Spaced Open Woodlands, Land Care Futures seminar:, Landscape Architecture Unit, Royal Melbourne Institute of Technology, quoted in R. Bird *et al.* 1992, 'The Role of Shelter for Protecting Soils, Plants and Livestock', *Agroforestry Systems*, Kluwer Academic Publishers, no. 20, pp. 59–86.

Knudtson, P. & Suzuki, D. 1992. *Wisdom of the Elders*, Allen & Unwin, Sydney.

Layson, A. n.d. Reverse Osmosis Applied to a Potable Water Supply—Coober Pedy (copy held by the author).

Long, J. 1971. 'Arid Region Aborigines: The Pintupi' in D. Mulvaney and J. Golson (eds), *Aboriginal Man in Australia*, Australian National University Press, Canberra.

Marriot, S., Nabben, T., Polkinghorne, L. & Youl, R. 1998, *Landcare in Australia—Founded on Local Action*, Landcare Australia, Melbourne.

Milton, D. J., Moss, F. J. & Barlow, B. C. 1978. *Regional Geology: Gosses Bluff Impact Structure Northern Territory, Scale 1:50,000, Plate 1*, Bureau of Mineral Resources, Geology & Geophysics, Canberra.

Mitchell, T. L. 1839. *Three Expeditions into the Interior of Eastern Australia . . .*, 2 vols, T. & W. Boone, London, quoted in R. Bird n.d., 'Trees in Western Victoria: An Historical Perspective', *Trees and Natural Resources*, vol. 28, no. 1, pp. 8–11.

Nathan, P. & Japanangka, D. L. 1983. *Settle Down Country: Pmere Arlaltyewele*, Kibble Books and Central Australian Aboriginal Congress, Malmsbury Vic.

New Zealand, Forest Research Institute, 1986. *What's New in Forest Research: High Quality Timber from Shelterbelts?*, Forest Research Institute, no. 141, Rotorua.

Pintupi Homelands Health Service 1994. Health in the Homelands—Spring 1994: A Report on the Health of Yanangu (unpublished).

Purvis, J. R. 1986. 'Nurture the Land: My Philosophies of Pastoral Management in Central Australia', *Australian Rangelands Journal*, vol. 8, no. 2, pp. 110–17.

Richards, D. 1995. 'Farmer of the Future', *Independent Monthly*, June.

Roe, P. 1983. *Gularabulu*, Fremantle Arts Centre Press, Fremantle WA.

Rose, D. B. 1996. *Nourishing Terrains: Australian Aboriginal Views of Landscape and Wilderness*, Australian Heritage Commission, Canberra.

Rowley, C. D. 1970. *Outcasts in Australia*, Australian National University Press, Canberra.

Ruddock, R. (comp.) 1995. *'Lanark' Farm Forestry*, 2nd edn, privately published, Melbourne.

Sinatra, J. B & Jones, D. S. (eds) 1988. 'RMIT Landscape Architecture Students Present: The "Lanark" Study', *Landscape Australia*, vol. 10, no. 4, pp. 349–77.

Sinatra, J. and Murphy, P. 1997. *Landscape for Health: Settlement Planning and Development for Better Health in Rural and Remote Indigenous Australia*, RMIT Outreach Australia Program, Melbourne.

South Australian Department of Mines and Energy 1986. *Opal in South Australia*, Mineral information series.

Turnball, J. W. (ed.) 1986. *Multipurpose Australian Trees and Shrubs: Lesser Known Species for Fuelwood and Agroforestry*, Australian Centre for International Agricultural Research, Canberra.

Victorian Archaeological Survey 1992. Lake Condah Aboriginal Sites: Heritage Significance Assessment, 525 1, Vic.

Victorian Department of Conservation and Natural Resources 1993. *Eastern Barred Bandicoot: Recovering an Endangered Species*, Information leaflet.

Water Resources Committee 1983. Preliminary Study for the Upgrading of the Coober Pedy Water Supply, Resources Committee, Coober Pedy Progress & Miners Association Inc.

Zwar, J. R., Beal, A. O. & Oddermatt, B. n.d. Water Efficient Public Landscaping in the S.A. Arid Zone (copy held by the author).